The Happiness of Pursuit

The Happiness of Pursuit

What Neuroscience Can Teach Us about the Good Life

SHIMON EDELMAN

BASIC BOOKS
A Member of the Perseus Books Group
New York

Published by Basic Books,
A Member of the Perseus Books Group

Books published by Basic Books are available at special
discounts for bulk purchases in the United States by
corporations, institutions, and other organizations. For more
information, please contact the Special Markets Department at
the Perseus Books Group, 2300 Chestnut Street, Suite 200,
Philadelphia, PA 19103, or call (800) 810-4145, ext. 5000, or
e-mail special.markets@perseusbooks.com.

Designed by Brent Wilcox

Edelman, Shimon.
    The happiness of pursuit : what neuroscience can teach us
about the good Life / Shimon Edelman.
        p. cm.
    Includes bibliographical references and index.
    ISBN 978-0-465-02224-3 (hardcover : alk. paper)
1. Thought and thinking.   2. Cognition.   3. Self.   I. Title.
    BF441.E234 2012
    153—dc23
                                                    2011039326

10 9 8 7 6 5 4 3 2 1

To Ira and Itamar—
Teach your parents well

CONTENTS

mustache and all, one step at a time. Where was I?
A moveable feast.

AUTHOR'S NOTE

According to one popular conception of science that goes all the way back to Francis Bacon's invention of it in 1620, scientific endeavor is all about getting answers from nature. That said, given that the quality of answers one gets depends conspicuously on the quality of the questions one asks, scientific inquiries lacking in intrepidity, imagination, and insight are likely to yield little more than scientifically validated tedium.

In the science of human existence, one half expects things to be easier. As someone who *lives* its subject matter, am I not in the best position to ask questions that go to the heart of it? And yet, it often seems to me that the really important questions hover at right angles to reality, manifesting themselves merely by a faint sense of unease or a premonition that I am about to miss the point of what is happening.

By acting swiftly and decisively, it is sometimes possible to apprehend a fleeting question and put it away for study. Here, I make a public example of four such questions, captured while stalking me on a hike through the canyon country of southern Utah:

1. A juniper tree, hanging on to a gravelly mound in a bend of the canyon, until the next flash flood.
2. A set of lizard tracks in the drying mud.
3. A dusty drive toward a far trailhead, down a narrow wash bordered by steep banks, arriving at length at an impassable sand trap.
4. A butterfly.

As you can see, such questions lose little of their cunning even in captivity, where they pretend that they are not questions at all, or that they are of no concern to the busy scientist and should be released into the custody of poets or philosophers. Such guile is best overcome by setting aside the conventional divisions between science and the humanities. This is why in this book my take on the life of the mind and how to make the most of it, while decidedly scientific, is not entirely conventional.

1 | Home Is Where the Mind Is

No justice, no peace.
A journey is mapped out.

Allons! whoever you are, come travel with me!
Traveling with me, you find what never tires.
—WALT WHITMAN,
Leaves of Grass: Song of the Open Road (1892, 82:9)

No Justice, No Peace

When I was eight years old, I read a book in which a few lines of a poem were quoted. The book was *Monday Starts on Saturday* by Arkady and Boris Strugatsky. The poem was by Christopher Logue, in a Russian translation. (I forgot to tell you that this was happening back in the USSR; the book's real title was *Понедельник начинается в субботу*.) The book, subtitled very aptly "A Fairy Tale for Younger Research Scientists," was about the daily life of applied magicians who work wonders by running experiments and solving equations. Whether or not this book contributed to my own eventual choice of career, I enjoyed it immensely. The poem, in contrast, must have gone right over my head—I have no recollection of it from that reading. As I discovered much later, it was an excerpt from Logue's "Epitaph":

You ask me:
What is the greatest happiness on earth? Two things:
changing my mind
as I change a penny for a shilling;
and
listening to the sound of a young girl
singing down the road
after she has asked me the way.

In subsequent rereadings of the book (once every few years ever since that first time), I have found myself becoming more and more intrigued by the poem. Maybe this is because happiness, with which it deals so deftly, is presented in the book as a challenge to the leading character, who works as the head of the computing center at the National Research Institute for Miracles and Magic. (By that time I was a computer scientist myself, albeit with very little magic and no miracles at all to my credit.) The challenge is implicit in a complaint voiced by an elderly graduate student, Magnus, who for decades now has been writing a dissertation for the Department of Linear Happiness and who offers Logue's poem as an example of the difficulties he faces:

Magnus sighed.
    —Some say one thing, others—another.
    —Tough,—I said with sympathy.

—Isn't it? How would you make sense of all this? To listen to the sound of a girl singing. . . . And not just any singing, but the girl is supposed to be young, down the road from him, and that too only after having asked him the way. . . . Is this any way to behave? As if such things could be algorithmized, huh?

For many years, I let those questions be. As a little boy, I was pretty happy not understanding algorithms, or girls. As a teenager, I was too busy trying to actually get a live girl to ask me the way. (As this was still in the USSR, I was unaware that the pursuit of happiness, along with life and liberty, is an inalienable human right, or I would have felt more relaxed about it.) As a computer scientist, I may have been mildly intrigued by whether or not happiness could be captured by an algorithm, but by then I had my own dissertation to worry about. Then I became a professor of psychology (a natural career move for a certain type of computer scientist . . . stick around and you'll see why), and things gradually took an unexpected turn.

I now had an excuse to think, on company time, about anything at all having to do with the human condition—a development that made me feel like a bear that wakes from hibernation to learn that a natural foods store specializing in bulk trail mix and artisanal honey has been built over its den. My research interests, which for many years had been confined to just a couple of the mind's faculties—mostly vision, then also language—began to broaden. Having discovered the same principles at work in both, I became curious about the rest.

By then, I was teaching a big introductory course on cognition, which, I felt, had to encompass everything that's known about how the mind works. Teaching, when taken seriously, does wonders to one's capacity for critical thinking; I realized that although the existing psychology textbooks were up to the moment on facts, they were decades behind on understanding. I ended up writing a text of my own, which I subtitled "How the Mind Really Works."[1]

For a while, the possibility of understanding things for myself with sufficient clarity to enable me to share my understanding with others made me vaguely happy. Then I perceived that the mandate that I claimed for myself came with a rider. If I truly grasped how the mind works, I should be able also to transcend all the usual vague intuitions about when, why, and how a person feels happy and replace them with sound scientific insight.

To my dismay, I realized that I would have no peace until the possibility of happiness being amenable to a scientific—perhaps even algorithmic—treatment was given, if not a decisive resolution, then at least a fair hearing. This book is my attempt at cajoling my conscience into letting me off that particular hook.

A Journey Is Mapped Out

To forestall the crushing skepticism that people tend to develop soon after hearing about someone embarking on this kind of project, let me explain why I think it is both timely and feasible. In the past several decades, tremendous progress has been made in understanding the mind/brain. It turns out that the principles that determine how the brain gives rise to the mind are very general, are statable in a pretty concise form, and have everything to do with computation. Given that the brain is the organ with which people experience happiness, understanding the brain offers for the first time a real chance for understanding how and why happiness happens, and perhaps for developing some recipes—algorithms!—for pursuing it more effectively.[2]

The focus on the *pursuit* of happiness, endorsed by the Declaration of Independence, fits well with the idea of life as a journey—a bright thread that runs through the literary canon of the collective human culture.[3] With the world at your feet, the turns that you should take along the way depend on what you are at the outset and on what you become as the journey lengthens. Accordingly, the present book is an attempt to understand, in a deeper sense than merely metaphorical, what it means to be human and how humans are shaped by the journey through this world, which the poet John Keats called "the vale of soul-making"—in particular, how it puts within the soul's reach "a bliss peculiar to each one's individual existence."[4]

The fundamental insight that serves as the starting point for my story is that the mind is inherently and essentially a bundle of ongoing computations, the brain being one of many possible substrates that can support them. I make the case for these claims by constructing, in plain sight and out of readily available materials, a conceptual toolbox that affords the reader a glimpse of the computations underlying the mind's faculties: perception, motivation and emotions, action, memory, thinking, social cognition, and language. This conceptual buildup culminates in an explanation that states, in plain language, the nature of the phenomenal self and of consciousness. Readers who are interested in the details that I omit can follow the leads offered by the many notes at the end of the book.[5]

These conceptual tools prove to be useful in making new sense of the notion of the pursuit of happiness. Quite satisfyingly, it emerges that the framers of the Declaration of Independence presaged the findings of the scientific inquiry into happiness: the dynamics of the self and of happiness is such that the pursuit it-self—the journey rather than the destination—is what really matters (hence the title of the book). This insight, such as it is, informs the book's conclusion: the seeker after happiness returns home, only to grow restless and eventually succumb to the lure of a new journey. On the basis of the understanding developed throughout the book, the following practical advice is offered as a way of summing up its lessons in seven words: when fishing for happiness, catch and release.

2 | Computing the Mind

A great metaphor that isn't. Concerning computation. No cognition without representation. Three things everyone should know about life, the universe, and everything. Promethean probabilities and amazing Bayes. Minds within brains. Minds without brains.

And if the body were not the Soul, what is the Soul?

—WALT WHITMAN,
Leaves of Grass: I Sing the Body Electric (19:1)

A Great Metaphor That Isn't

Let me tell you a short story about the brain. The brain is the most complex object known to science. Because scientists have been unable to explain exactly how the brain gives rise to the mind, they keep resorting to technological metaphors of complexity. The best metaphors are those that draw on concepts associated with newfangled technologies, which still exude a certain aura of mystery. In the past, such metaphors came from mechanics ("the brain is an intricate clockwork") and electronics ("the brain is a vast telephone exchange"). For some time now, everyone's favorite source of metaphors has been computer engineering. Still, no matter how much we like to compare brains ("meat computers") and computers ("electronic brains"), the computer metaphor is merely the latest installment in a long series of fads to which brain science periodically succumbs.

If you find yourself liking this story, you are in good company. The "computer metaphor" view of how the mind works sounds sophisticated and modest at the same time and is particularly intellectually appealing to progressive-minded people who know their history of science and value openness toward the prospect of continued replacement of good theories by better ones. It is also popular with science writers, including those practicing cognitive scientists who are eager to share their findings and insights with anyone who is interested in how the mind works. For them, the computer metaphor offers a neat way to introduce and explain the tremendous progress made by cognitive science in the three decades since computational theorizing first started to prove uncannily effective. Better yet, they can do so without actually calling the reader a computing machine.

But what if you really are one?

In our daily lives we routinely encounter devices that can only be understood in terms of computation. Take grocery-store cash registers as an example. These come in different sizes and colors and may rely on diverse mechanical and electronic components, but they all have one inalienable, categorical, defining feature in common: they compute. Take away a cash register's ability to compute and you're left with a heap of junk, the machine equivalent of a dead body. In the deepest possible sense, computing is what cash registers are fundamentally about. It would be intellectually irresponsible to insist that doing sums is only one among many equally valid ways of describing the function of a cash register or to argue that a cash register is only metaphorically a computing machine.

This observation sets the stage for the unveiling of what is undoubtedly both the most important and the least-kept secret of cognitive science. Although it has entire books devoted to it, this secret has so far managed to elude the attention of most of the general public and even of some cognitive scientists. It has been able to hide in plain sight because of its revolutionary implications. (We humans often blissfully ignore inconvenient truths, even as we stare them in the face.) Here's the secret, then: computation is just as much a defining feature of brains as it is of cash registers. Moreover, in both cases it is the most important such characteristic: a cash register's very existence (let alone its mechanics or electronics) can be really understood only by resorting to the concept of computation, and so can the brain's.[1]

Because it takes a while to do justice to the idea that cognition is computation, let me offer you right away a few quick examples that illustrate it. My first example is as simple as black on white—the black of the print against the white of the paper in front of you. Just as all cats are black in the dark, both the paper and the print would be equally invisible if it were not for the light that illuminates the page. Intuitively, the paper reflects more of the light that falls on it than the print does, which is why the letters are seen as darker than their background. It would seem, therefore, that telling apart the print and the paper boils down to gauging the amount of light reflected from each. There is, however, a complication: the amount of light that enters your eye after hitting the page is determined by two independent factors: the quantity of light that is available to begin with (the intensity of the illumination) and the fraction that is reflected from the page (its reflectance).

To fully appreciate the challenge faced (and met!) by your visual system, even as you are reading these lines, let's state it in concrete and precise terms. (It's okay to skip the numerical example that follows if you already saw the light.) Suppose your eye is registering 100 photons per second arriving from the region of the page that you are looking at. This measurement can result from various combinations of illumination and page reflectance: for example, 10,000 photons (strong illumination) falling on it with only 1 percent of them being reflected back (low-reflectance or "dark" surface), or 125 photons (weak illumination) falling, of which 80 percent are reflected (high-reflectance or "light" surface). A quick reflection (do try it at home!) reveals that there is an infinity of possible pairings of illumination and surface reflectance values that can give rise to the very same number of photons reaching the eye. Which pairing is the right one?

To find that out, the brain must solve a problem that is fundamentally computational: given a product of two numbers (illumination and reflectance), determine what they are individually. Your ability to perceive the ink as black and the paper as white in direct sunlight as well as in deep shade is clear evidence that your brain indeed manages to solve this inherently arithmetical problem. How it does that is beside the point for the moment.[2] Let's just admit, as we must, that at least one everyday perceptual task can only be solved through computation, because this task cannot even be stated without resorting to numbers (which is why I had no choice but to mention numbers while introducing it just now).

My second example involves the task of thinking. Intuitively, thinking is what follows perception (sizing up the situation you're in) and precedes action (doing something about it). Humans are pretty good at abstract thinking (witness the ability of some of us to stand up to chess-playing computers, sometimes for as long as a couple dozen moves), but it would be more useful for us to focus here on more mundane challenges, such as figuring out which register line in a supermarket checkout area to join. The simplest approach to this problem, which I personally face much more often than I play chess, is to estimate the length of each line in person-units and then to join the shortest one. Those of us with some supermarket experience are likely to see the simple head-counting approach as unsatisfactory. We know that a better estimate of the amount of time one is likely to spend standing in a line depends not only on the number of people ahead of you, but also on the number of items each of them is buying and on the efficiency of the cashier, as measured by the number of items he or she can scan and bag per minute.

This example illustrates nicely the value of thinking before doing: a few seconds spent on observing a line and thinking about what you see can save you quite a few minutes' worth of waiting time, which seems like a reasonable return on investment. During these seconds, your brain forms estimates of the number of people in line and the number of items each one has, multiplies these two numbers together, and divides the product by an estimate of the cashier's efficiency. Do it for each line you're considering, and you have the proper grounds for making an informed choice about the most promising line. Thus, at least some problems that require thinking, just like perception, reduce to the manipulation of numbers according to certain rules, in this case multiplication, division, and comparison.

My third and for now last example has to do with a drinking problem, albeit not of the kind that necessarily involves alcohol. This particular problem arises in the planning of bodily movements. Imagine yourself sitting at a dining table, with your hand around a glass of water that rests in front of you. You are thirsty, but before you can quench your thirst your brain must crunch some numbers. Because the incident that we are imagining is set in the three-dimensional space of the dining room, it takes three numbers to specify the location of the hand that is holding the glass. If you single out one corner of the room and measure how far your hand is from each of the three surfaces (two walls and the floor) that meet at that corner, you can pinpoint its location precisely. As far as your brain is concerned, however, many more than three numbers are needed. This is because the brain does not measure or control directly the distances between your hand and the walls.

What the brain controls is the angles of the various joints of the body, of which there are many. Looking just at your upper extremities, you can count three independently controllable angles at the shoulder (direction in the horizontal plane, direction in the vertical plane, and rotation), two at the elbow, and two at the wrist. So the very formulation of the problem of planning how to get your hand with the glass from its resting position on the table to your mouth involves a whole series of pretty scary-sounding computations. First, the brain must establish a correspondence between locations specified in the "room" format (as triplets of numbers) and the same locations specified in the "body" format (as lists of seven numbers). Second, it must use this correspondence to compute the seven-number setting that would bring your hand with the glass to your mouth.

Formidable as it is, this is merely a simplified version of the actual, full-blown problem of hand movement control, which requires more than bringing the glass to your mouth (rather than, say, to your ear). For one thing, you would probably like to be able to do it without spilling the water along the way or smashing in your teeth at the end. Let me set these complications aside and reiterate instead the key point: in the three examples offered here, the problems that arise, which I chose from the three main areas of cognition—perception, action, and what's in between—are all inherently computational. These problems, which the brain encounters and solves as a matter of daily routine and usually outside of conscious deliberation, cannot even be stated without recourse to numbers, and their solution must therefore involve some kind of number crunching.

This realization spells a certain kind of doom for the venerable computer metaphor for the brain, with which I opened this chapter. In science, a metaphor may wither away after being made irrelevant by new insights into the phenomenon that motivated it in the first place. Alternatively, it may crystallize into an accepted explanation, if theoretical advances and empirical findings vindicate it. Such is the fate of the computer metaphor in brain science—a truly great metaphor that isn't.

Concerning Computation

Given that the computer metaphor for the brain looks like it has the embarrassing (for a metaphor) quality of being literally true, we must make sure that we understand what computation is, if we ever want to be able to fully understand brains. Luckily, understanding the basic nature of computation does not require a degree in computer science. Contrary to what one is inclined to believe, most instances of computation in the universe happen without involving any kind of particularly complex, contrived, or specially engineered device. Computation is all around us; in fact, it is as common as anything can be, for the following simple reason: every physical process that unfolds over time computes something.

Think about it. How would a pebble that you release from your grasp be able to strike the ground precisely at the correct instant if it could not figure out—compute—its trajectory, given the acceleration of gravity and its initial position? How would a stove-top heating element be able to reach precisely the correct temperature if it could not figure out the current that flows through it, given the grid's voltage and its own resistance? In a very concrete sense, a falling pebble computes its trajectory by following Newton's laws, just as a heating element computes the current that flows through it by following Ohm's law.

There is of course nothing mysterious about objects obeying the laws of physics: on the contrary, it would be a mystery (in other words, a sign of a subtler law at work) if they did not, or if exceptions or exemptions were possible. This situation seems to be not to everyone's liking: riding a ski chairlift one winter day, I saw on a pylon an ad that began: GRAVITY: IT'S MORE OF A NUISANCE THAN A LAW—a bizarre notion, given that not just skiing but actually getting down at all from the mountain and seeing your loved ones ever again (rather than flying off screaming into space) would be impossible, were it not for gravity. Like it or not, the implacability of the laws of physics is what makes the world go round, so to speak, and in doing so carry out a vast panoply of computations.

Most of these computations are very limited in the spatial and temporal reach of their effects. Such is the fate of the computation performed by the pebble that I release and let fall to the ground. The law-abiding pebble faithfully computes its own trajectory and manages to arrive at its resting place at a precisely timed moment, yet this computation is pretty inconsequential: a few grains of sand may be kicked around, but once they settle the only readily discernible change in the state of affairs of the universe is that the pebble, which was previously in my hand, is now on the ground. Who cares?

That particular computation was useless, but it could have been otherwise, if only the fall of the pebble were to have some significant and enduring consequences or repercussions on other events. Here is how this could happen. Imagine that you are defending a castle. Your assigned post is immediately above the portcullis, and your duty is to drop an anvil onto the head of any attacker who comes too close to it. Your store of ammunition is limited (the besiegers just intercepted an inbound shipment of blacksmith supplies), and so you would like to time the release of each anvil so that it arrives at a certain small volume of space simultaneously with the head of the intended target. This can be easily arranged through practice with some pebbles. All you need to do is mark the position of the knight who is charging at the gate as you release the pebble; if the pebble then strikes the charging knight's helmet (you'll hear it), you can be reasonably sure that the next time a knight's charge takes him through the marked position you may drop the actual anvil instead of the practice pebble.

You may be surprised to realize that what I just described happens to be a computer- controlled ballistic missile launch system. The missile is, of course, the anvil, which is ballistic by definition, because it has no onboard means of propulsion. And what about the launch-control computer? Since you are in charge of releasing and observing the pebble, you are one of its parts. Another part is the earth, whose gravity field so conveniently imparts an equal acceleration to anvils and to pebbles (as Galileo famously discovered four hundred years ago). Yet another part is the pebble that you use to simulate the anvil's projected descent. Unlike the listless, lonely pebble from the earlier example, about whose fall nobody cares, this one is lovingly tracked and the computation it carries out (just by being itself and acting naturally) is made use of. Its use lies in serving as a stand-in for or a *representation* of the anvil: the actual fall of the pebble, as it unfolds over time, represents the potential descent of the anvil.

Because the fall of the pebble is precisely analogous to the fall of the anvil that it represents (their trajectories unfold at exactly the same rate), the anvil-aiming computer of which the pebble is part is an *analog* computer. The familiar *digital* computers differ from analog ones in one key respect: they deal in representations that are made to correspond to their objects through some arbitrary but consistent rules rather than through direct physical analogy. A good example of a digital representation is the code agreed upon in April 1775 by Paul Revere and Robert Newman, the sexton of the Old North Church in Boston, for representing the mode of the expected British assault: one if by land, two if by sea.

There is nothing inherent in assault by land that dictates that it should be represented by one lantern in the belfry as opposed to two. It is in this sense that the Revere-Newman code is symbolic and arbitrary. Switching to a more modern example, consider how my notebook's battery meter represents the remaining charge. If I have worked for an hour since the last full charge and the meter now shows "75%," I conclude that the battery will last for three more hours. The connection between the symbols "75%" and what I take them to represent is entirely a matter of convention; the same information could have been equally well represented by the symbols "3/4." Alternatively, I can switch the battery meter from the digital representation mode to analog, in which case it would look like a little "thermometer" whose mercury is three-quarters of the way up to the maximum mark.

What is common to all these examples of representation is that some *physical symbol*—an object, event, or process—stands in for some other object, event, or process for the benefit of a third party. Thus, some squiggles on my notebook screen represent for me the battery level; the appearance of a lantern in a belfry conveys to Paul Revere a piece of battle intelligence; the fall of a pebble tells the castle defender how the fall of an anvil would proceed. Understanding representation gives us a crucially important conceptual handle on computation. The key insight is this: useful computation hinges on the possibility of some objects or processes representing others. Indeed, the reliance on representations is the defining characteristic of useful computation, which distinguishes it from computation that merely happens as the world goes on.

No Cognition Without Representation

Against the constant, pervasive background hum of the universe relentlessly computing its next state given its past history, the relationship of representation—two computations that always co-occur in a particular order or unfold over time in lockstep with one another—stands out as a great rarity. Merely drawing a parallel between some carefully chosen aspects of two events is not at all difficult, especially if it is numerical or otherwise abstract; finding a repeated co-occurrence or a parallel that persists over time is.

As I ponder this point, I observe nine ducks landing on a pond and am struck by the thought that their number is exactly the same as the number of months I have left until the date on which

I promised to deliver this book to the publisher. What am I to make of this observation? It is this kind of coincidence, invariably noted in retrospect, that inspires popular tales of successful auguries, such as the one found in the second book of *The Iliad*. The Greeks, weather-bound at Aulis on their way to Troy, witness an omen: a snake devours a mother sparrow and its eight chicks in their nest, then turns into stone. They are told by Calchas, their chief soothsayer, that the omen means that the coming war with Troy will not be resolved until the tenth year. It is worth noting that this story betrays perfect hindsight on Homer's part: the prophesied tenth year happens to be the very year during which the main events of *The Iliad* are set.[3]

The strange decision to count together the sparrows and the newly petrified snake to reach the number ten would have ruined the career of a lesser augur than Calchas (who did not seem to have had any problem convincing the Greeks at Aulis that King Agamemnon's daughter had to be sacrificed to Artemis to break the spell of bad weather). Scholars believe that the sparrow omen story confounded Aristotle, who was otherwise a great admirer of *The Iliad*.[4] Ever since, educated people's readiness to take such omens at face value has been steadily on the wane. Indeed, some people viewed numerological divination with great suspicion already in biblical times, as suggested by the injunction found in Leviticus: "Nor shall ye employ auguries, nor divine by inspection of birds." One likes to think that the Old Testament lawgivers sought to protect the people from superstitiously perceiving false patterns in noise (rather than merely to protect their own monopoly on foretelling).

The more complex a set of observations, the more likely it is to give rise to ghost "patterns," or correlations that arise by chance—an apparent signal in a sea of random noise. Skepticism toward such correlations is the only reasonable default stance in science, and it can only be assuaged through the application of science's single most powerful ghost-busting tool: statistical analysis of data. Patterns that withstand statistical scrutiny are still suspect: they need to be *explained*, that is, integrated into a wider framework or theory that makes sense of as much data as possible, in the simplest possible terms. This is how observations (themselves usually guided by informed guesses or hypotheses) get distilled into laws of nature.

Now that we are armed with the scientific method, let us return to the problem of telling apart reliable from unreliable representations—that is, true co-occurrences between events from chance ones. The first hypothesis to be let go is the one stating that a bunch of objects represents another bunch of different objects simply because their number is the same. As if to prove this point, my nine ducks on a pond are joined by another; I rush home and open my mail to see if the deadline for my manuscript has receded by a month, only to find out that, alas, it is still only nine months away. Apparently, the number of ducks on that particular pond does not, after all, represent the time frame of my contract with the publisher. More generally, attaching wide-ranging representational significance to patterns of bird flocking is empirically unwarranted, as one can quickly learn by observing the efficacy of the hypothesized representations over time.

It would be even better if we could tell ahead of time, from first principles, whether or not one thing—object, event, process— is going to be representative of another. Thanks to the understanding of the natural world that has already been gained by science, it is possible to do so. What is it then that singles out valid instances of representation? There is just one principle that can ensure the validity of a representation: *causation*, as it is embodied in the various laws of nature.[5] The fall of a pebble and the fall of an anvil are caused and governed instant by instant by the same universal law—the law that Newton is said to have discovered when he intuited that the motion of the celestial bodies is not any different from the motion of terrestrial ones. Because of that common causality, falling pebbles are representative of falling anvils (and apples and penguins, but not ducks). We may think of the relation between processes that all obey the same mathematical equation as representation by causally justified analogy.

There is another way in which causality can give rise to a valid representational relation: one event may simply cause the other. This is how the lighting of a lantern in the belfry of a church on the eve of the Battle of Lexington and Concord represented the redcoats' advance (through the mediation of one Robert Newman). Representation that is underwritten by causality in this direct manner is much more versatile than causal analogy: whereas falling objects representative of each other must all resemble each other in certain respects (heft, flightlessness), smoke represents fire by virtue of being caused by it, not by resembling it. Moreover, it does not matter how far from the represented event the causal chain leads before it reaches the representing event, as long as the chain is reliable; thus, a blaring fire alarm represents fire even if it is activated by smoke.

The principle of representation of one event by another through a causal chain whose reliability is guaranteed by the laws of nature is extremely powerful. Once it is discovered by evolution[6] and unleashed into the biosphere, it becomes a force that can change the face of the planet. Whereas computation as such permeates the very fabric of the universe, only those rare physical systems that harbor representations of their environment are capable of sustaining *cognitive* computation, whose distinctive feature is that it is about something other than itself. In a typical cognitive system, the representing processes are separated from the represented ones by being safely ensconced inside a contraption that keeps the rest of the world at bay while sensing its surroundings, computing a course of action, and carrying it out, all at the same time. The most familiar naturally occurring contraption of that kind is an embodied brain.

Three Things Everyone Should Know About Life, the Universe, and Everything

The driving force and the arbiter of innovation in biology is evolution, and so a species that invests resources in representing its environment will thrive only if computing with its representations makes it, in the final account, a better replicator. The key question that needs to be asked about brain computation is therefore one of utility: what survival and reproduction advantages does having minds—which is what brains compute—confer on animals?

It may seem that the answer to this question would depend a lot on the animal, on where it lives, and on the company it keeps—but it does not. Instead, three basic facts of life, the universe, and everything join forces in creating a simple and general explanation for the myriad ways in which mindful hominids, squirrels, chickadees, or octopuses gain an upper hand, paw, foot, or tentacle over happy-go-lucky ones.

The first basic fact is about the universe. It so happens that the universe we inhabit has built into it an asymmetry between the past and the future. The arrow of time, which is what this asymmetry is often called, emerges independently from the action of several distinct families of laws of physics. Among these are the laws of thermodynamics, which govern the fundamentals of self-organization, stability, and self-replication—in other words, the origins, maintenance, and propagation of life.

The predicament of being alive is what the second basic fact is about. Life is fragile and time is irreversible. An antelope that sees its life passing in front of it as it is being brought down by a lion cannot roll the nature documentary in which it is starring back to that critical moment, only seconds before, when it ignored the flicker of movement in the tall grass. An animal that gets eaten or dies of starvation or is run over by a car or sets itself on fire by playing with matches is gone forever. The only effective insurance against such outcomes is to recognize the threats before it is too late.

Threats are always about the future, whereas the experience on which we can base our actions is all in the past. What is a poor animal to do? Seeing that antelopes, lions, and humans are still around, it must be possible to evade or otherwise thwart at least some impending disasters. This brings us to the third basic fact, which is about everything—that is, everything that's relevant to the life of an animal. The fact is this: the future state of affairs in an ecological niche is predictable from its past, up to a point.

The reason for this predictability is the sheer physical inertia of the universe. On an appropriately short time scale, things are guaranteed to stay as they are or to carry on changing in the same manner as they did before. There are also many kinds of longer-term regularities, such as cycles of seasons. Animals can evolve to rely upon seasonal changes in the environment and to anticipate them from telltale cues (think of migratory birds that respond to the first frost). Or, animals can evolve sophisticated brains that represent patterns of change in the environment and anticipate the future by treating it as a statistically projected extension of the past. Either way, it is the capacity for *forethought* that distinguishes, on the average, between the quick and the dead.

What are brains good for, then, and what preeminence doth a beast with a better brain have over others? As I promised, the answer is short and simple: it's all about forethought. This conclusion hinges on the three key facts about life, the universe, and everything. Forethought is meaningful, important, and practical because time is directional, because it pays to anticipate the future, and because it is possible to do so by consulting the past.

The capacity for forethought or foresight sounds so advanced that attributing it to beasts may seem like a big stretch. In ancient Greek beliefs, forethought elevated men above beasts. According to the creation myth retold by Hesiod, Prometheus, the Titan whose name *means* "forethought," stole fire from Zeus and gave it, along with the ability to reason and sundry other survival skills, to men (but not to women, whom Zeus created later, to balance the good brought to the world of men by Prometheus). Israelite sages held foresight in even greater esteem: the Talmud asks rhetorically, "Whosoever is wise?" and replies, "The one who sees that which is about to be born." (In the spirit of the times, those wise men too would have laughed you out of the shul for suggesting that women have sense.)[7]

In reality, forethought is found in the lowliest places on earth, and even underneath it. My favorite example illustrating that you don't have to be male, human, or even sighted to possess a modicum of foresight involves blind mole rats (of either sex). If someone digs a ditch overnight across the path to your front door, you will most likely see it in time to avoid falling in. For that, you need only exercise foresight literally and passively, by simply looking ahead. In comparison, the humble mole rat,

Spalax ehrenbergi, exercises foresight metaphorically (necessarily, because it is blind) and actively. Here's how it works: a burrowing mole rat bangs its head against the ceiling of the tunnel and analyzes the returning echo for signs of obstacles such as rocks or ditches.[8]

This feat of subterranean echolocation was discovered only recently, in part because researchers have tried for years to trap mole rats for study by waiting for them to fall into freshly dug ditches. It does quite raise the ante on the human bid for sophistication in seeing the shape of things to come. How are we to make sense of it? It so happens that foresight and forethought are evolvable. Having a better brain—one that computes a sharper mind—can make the life of its owner longer, or at least long enough for it to mate. Thus, evolution generally favors creatures that are being mindful about the future—whether by casting their brains about or by banging their head against the ceiling of the burrow—over the happy-go-lucky ones that are oblivious to what the future may bring to their piece of turf.[9]

Promethean Probabilities and Amazing Bayes

Forethought is a wonderfully versatile tool: when properly tuned and regularly consulted, it safeguards your life and liberty, whether you are a burrowing rodent or a homeowner whose front lawn has been replaced overnight by a newly dug ditch. It also promotes the pursuit of happiness, both in a big way (after all, evolution hones it to catalyze mating, which is often fun) and in various small ways, as when a grocery shopper chooses the checkout line that he or she projects to result in the shortest and therefore happiest waiting experience. Given that the future is predictable to the extent that the world exhibits statistical regularities, how do animal brains compute it? The same way statisticians do it, in all likelihood.

Brains end up resorting to statistics for the same two reasons that governments and corporate managers do so: first, because there is always more potentially relevant raw data available than can be effectively grasped; and second, because the available data are uncertain. Amassing the kind of descriptive statistics usually reported in popular media, such as the means of some quantities of interest (say, the mean income per capita), does not address the really interesting questions arising from the data, namely, what it could all possibly mean and where things may be heading. Thus, it is the second reason that makes statistics vitally important.

Unlike descriptive statistics, which at best merely quantifies uncertainty, *inferential statistics* puts it to work by processing available data in search of trends or patterns on which predictions about future outcomes can be based. Although a hypothesis-driven search of data for patterns is a standard operating procedure of any empirical science, you don't have to be a scientist to make it happen. In fact, just as with foresight, you don't even have to be human.

Consider the case of an adolescent blind mole rat, whom I shall call Molly (not her real name). Molly has just decided to reject the entrenched ways of her parents and is about to start digging at right angles to her home tunnel. What does she expect to find there? Soft earth, a rock face, a ditch? Her beliefs in this matter are simple and undiscriminating: she holds these three possibilities to be equally likely.

Molly may be blind and inexperienced, but she is neither senseless nor stupid. Rather than clinging obstinately to the a priori hypothesis, she performs, as she digs on, occasional experiments to gather new data (by banging her head against the tunnel ceiling, of course—the one useful skill she admits to having inherited from her family). She processes the echo from each head-slam to glean some information about the way ahead.

As she adds this new information to the existing pool of data, one of the possibilities grows more likely and the others less so. Soon little doubt is left that the way ahead leads into an open space, but Molly wants to feel it with her own whiskers. After digging a bit further, she feels on her face the draft of fresh air that all mole rats instinctively shun. With a shrug and a sigh, Molly turns back.

Let us see how Molly's actions allow her to learn from experience and to become better at using her echo sense to feel her way ahead. The quantities of interest are the relative likelihoods of the various possible outcomes (ditch, rock, earth). These relative likelihoods are simply the ratios of the corresponding subjective probabilities; we may think of these probabilities as expressing Molly's beliefs about various possible states of affairs in her subterranean world. The prior probabilities are all the same (one third each for ditch, rock, earth), and so the likelihood ratios are all equal to one.

Once new echo information from a head-banging experiment comes in, it must be taken into account. What Molly needs to compute, then, is the *conditional probability* of each of the possible outcomes: the probability of ditch *given* the particulars of the echo (and likewise for rock and earth). Intuitively, this conditional probability expresses the degree to which a belief (that what lies ahead is ditch) must be modified by additional information (that what lies ahead returns a particular kind of echo). How can it be computed?

It turns out that the unknown conditional probability that Molly is after is proportional to the product of two known quantities: the prior probability of ditch and the conditional probability of the echo given ditch. The first factor is easy (it is simply the belief that Molly holds before the new data are in), but what about the second one? It can be estimated if Molly keeps track of actual outcomes that follow each kind of echo. For instance, an encounter with a ditch that follows the experience of a particular echo can contribute to the estimate of the probability of that echo, given that there is a ditch there; and while she's at it, Molly can also update her cumulative estimate of the prior (unconditional) probability of ditch.

The mathematical foundations for updating subjective probabilities and for managing uncertainty by incorporating new empirical data into a classification or decision procedure were laid by an eighteenth-century amateur statistician, Thomas Bayes. (This makes Molly a Bayesian, even though she does not know it.) The key contribution on the part of Bayes was a theorem that expresses the so-called posterior probability (posterior to the experiment, that is; in our example this is the conditional probability of ditch given echo) in terms of the product of the prior (the probability of ditch before the head-bang) and the so-called data likelihood term (the conditional probability of the echo given ditch).[10]

I decided against including the proof of the Bayes Theorem here, even though it is only three lines long. The received wisdom is that every equation in a book drives away half of the readers, and I do not want to risk losing seven-eighths of my readership. (Come to think of it, this disclaimer alone may be too quantitative for its own good.) I'll just go on treating the Bayes Theorem as a miracle then.

Considering how useful Bayesian statistics is all across cognition, it certainly deserves to be called a miracle. In perception, the Bayesian formula for posterior probability comes in handy for updating the representation of the environment given the sensory data. In decision making, it is indispensable for taking into account new evidence while choosing the best outcome. In action, it is used for planning a sequence of motor commands that combine the current goal with prior experience.

All these cognitive functions are implicitly included in the example of Molly the mole rat, and all of them ultimately serve the supreme goal of cognition: forethought, which we are now in a position to understand as processing past experience so as to be able to deal more effectively with the future. Because prediction is tricky (especially, they say, about the future), principled management of uncertainty is absolutely crucial in this undertaking. The Bayesian framework for statistical inference and decision making is exceptionally well suited for meeting this challenge. As we will see next, it is also well suited for being implemented with the standard building blocks of brains: nerve cells, or neurons.

Minds Within Brains

Having found out that minds are for forethought and that forethought can be distilled from experience using statistical inference, we can now begin to understand how brains compute minds. As we know, brains compute useful things for their owners by harboring representations of the world. Generally speaking, they do it by maintaining *internal states*, which stand for various aspects of the external world in a consistent manner that captures their causal structure. The state of a brain is a fleeting, dynamical thing: being simply the activities of all the neurons of which the brain consists, the state changes from one instant to the next, driven by its past history and by external inputs.

In being dynamical, a brain does not differ from any other physical-law-abiding system, such as the pebble-based missile control computer that I described earlier in this chapter. (I consider this good news: the more humdrum the emerging explanation of the mind is, the easier it is to relate to.) Even so, the dynamics of a brain, especially a large one such as yours or mine, is much, much more involved—because neurons, unlike pebbles or anvils, are a piece of work.

First, neural representations are *active*. A pebble resting on the ground after a fall can represent a fallen anvil (boring); if you need it to represent a *falling* anvil, you'll have to pick it up first. In comparison, neurons pick themselves up every time. A neuron that just finished representing a bit of outside information, which it does by sending a voltage spike down its output fiber or axon, becomes ready to do so all over again under its own power after just a few thousandths of a second. All it asks for in return is some glucose and some oxygen.

Second, neurons *network*. Axons make connections to other neurons at junctions that are called *synapses*. Each of these acts like a throttle that controls the strength of the tiny kick of electrical current imparted to the target neuron by an incoming spike. Each of the brain's billions of neurons connects to others, whose numbers may run into the tens of thousands. Large-scale networking is no longer the exclusive province of nervous systems that it was before the Internet became a household word, and yet there is no artificial network out there that packs so much knottiness into such a small volume as the human brain.

Third, neural activity is *patterned*. The cumulative effect of many incoming kicks may push the target neuron over the brink, causing it to fire. Because their individual contributions are typically very small, spikes from many source neurons must converge on the target neuron within a short time of each other for it to fire. Moreover, because each spike's kick is regulated by the efficacy or "weight" of the synapse through which it is delivered, only certain specific coalitions of other neurons can set off a given neuron. Thus, neural activity is highly structured both in time and in space.

Fourth, neurons *learn from experience*. Much of this learning takes the form of activity-dependent modification of the hundreds of billions of synapses connecting neurons to each other. In this type of learning, a synapse is made a bit stronger every time it delivers a kick just before its target neuron fires and a bit weaker every time the kick arrives slightly too late to make a difference. Mathematical analysis and empirical investigations show that this simple rule for synaptic modification can cause ensembles of neurons to self-organize into performing certain kinds of statistical inference on their inputs, thereby learning representations that support Bayesian decision making.

To see how statistical computation can be carried out by brains, think of a neuron in the brain of a mole rat that comes to represent the presence of a ditch within her tactile sensory range. If this neuron's axon connects to another neuron, which has learned to represent the sensory quality of the echo that the mole-rat experienced shortly beforehand, then the weight of the synapse between them can be seen to represent the conditional probability of the echo, given the presence of the ditch—one of the quantities that the Bayesian brain needs to exercise foresight.

It turns out that a network of neurons is a natural biological medium for representing a network of causal relationships. Within this medium, the activities of neurons stand for objects or events, and their connections represent the patterns of causation: conditional probabilities, context, exceptions, and so on.

Being inherently numerical, the currency that neurons trade with one another—the numbers of spikes they emit and their timing—is the most versatile kind of physical symbol. By being able to learn and use numerical representations, neurons leave pebbles and anvils (considered as physical symbols) in the dust. Of course, any symbol, numerical or not, can stand for anything at all. However, numerical symbols are absolutely required if the representational system needs to deal with *quantities* that must be mathematically relatable to each other, as in the probabilistic computation of causal knowledge.

Thus, not only are networks of neurons exquisitely suitable for representing the world and making statistically grounded foresight possible, but they can learn to do so on their own, as synapses change in response to the relative strength and timing of the activities of the neurons they connect. Seeing neural computation in this light goes a long way toward demystifying the role of the brain in making its owner be mindful of the world at large. The collective doings of the brain's multitudes of neurons may be mind-boggling to contemplate, but that's only because ex- planatory value—that is, conceptual simplicity—is found in the principles, not the details, of what the brain does.

Minds Without Brains

One of my favorite concise descriptions of the nature of the human mind comes from mathematician and computer scientist Marvin Minsky, who once observed that the mind is what the brain does. Having gotten a glimpse of the principles of what the mind is (a bundle of computations in the service of forethought) and of what the brain does (carrying out those computations), we can appreciate Minsky's quip, but also discern that it is open to a very intriguing interpretation. The point of it is this: if what the brain does can be done by other means, then a mind can arise without the need for a brain.

To make peace with this outrageous yet true proposition, we need to focus on a key characteristic of computation: identical computations can be carried out by radically different physical means. This characteristic figured already in the very first example that I used to introduce the concept of computation in this book—that of the pebble and the anvil. Each of these two objects computes, by analog means, virtually the same thing (a particular kind of trajectory that must be followed while falling down to earth), which is why a pebble can represent the anvil (and vice versa).

Intuition suggests that this very same analog computation can be carried out by throwing any of a number of other objects—such as penguins (as I noted earlier in this chapter) or, to pick a less trite example, marmots. It appears that we may change many of the object's properties without altering at all the computation that it carries out by undergoing the process of falling. For instance, we may vary freely the ferocity of the thrown animal: falling ferrets and rabbits would do equally well in computing the trajectory of a falling anvil. All this is so because pebbles, anvils, penguins, marmots, ferrets, and rabbits share the one physical feature that is absolutely required for the analog computation in question, namely, a high ratio of mass to air resistance.

What special physical features are required for computing a mind? None at all! That networks of neurons and learning synapses are exceedingly good at their job does not preclude mind-like computation from being enacted by some other kind of contraption. Remember that the brain's neurons compute the mind by multiplying, adding, and passing around *numbers* that stand for various aspects of the world and of the brain's own internal states, and numbers don't care what they are made of. For a sorting machine whose function is to count roughly fist-sized fruit, nine apples means exactly the same thing as nine oranges. More generally, thoroughly different machines can be made to compute exactly the same thing: an 1884 vintage mechanical cash register is just as good at doing sums as an application that emulates it, complete with the clanking and the concluding bell chime, on an early-twenty-first-century handheld computing device.

This implies that any machine that can carry out a particular brain's number game—as it unfolds over time and down to the last bit—would give rise to precisely the same mind that the brain does.[11] This equality has one consequence that only the most bigoted computing machine would fail to appreciate: in the society of minds, it does not matter what you compute yourself with or whether your grandmother had gears, vacuum tubes, or actual blood-soaked gray goo in her brainpan. On top of that, the mind is a moveable feast: although the original home of the human mind is in the human brain, it can flourish in any other medium that does faithfully and well enough what the brain does so well.

✍ SYNOPSIS

The familiar "computer metaphor" that halfheartedly likens the brain to a computer must be discarded: it is unnecessary and in fact inappropriate, because the mind is computational in a literal sense. It is easy to explain the concept of computation in plain terms: it turns out that every physical process computes something. Which physical processes are cognitive processes? Those that operate on representations—internal stand-ins for objects and events that are external to the system in question.

To establish the relevance of computation to cognition, we need to consider examples of perceptual, motor, and other tasks that can only be solved by crunching numbers, some of which stand for various entities external to the brain and others for its internal states. It turns out that all of the mind's tasks are like that. More generally, minds evolved to support foresight, which brains compute by learning and using the statistics of the world in which they live.

Thus, the mind is best defined as the bundle of computations carried out collectively by the brain's neurons. Because the same computation can be implemented by different physical means, nonbiological minds are revealed as a distinct possibility. This line of reasoning, supported by hard evidence from cognitive sciences, exposes the mind-body problem as an artifact of old ways of conceptualizing cognition that can be safely dismissed.

3 | The Republic of Soul

A discourse on method. Faster than a speeding marmot. A treatise of human nature. Perception by numbers. Representation space: the final frontier. Being in the world. The instruments of change. The value of everything. Things get interesting.

Between the motion
And the act
Falls the Shadow.

    —T. S. ELIOT, *The Hollow Men* (1925)

. . . at this moment, I'd say, I am
a bringer of light; a man who stands in a doorway
flooded by sun;
I am a bird; someone who learns,
in shadow, the real shape of brightness.

    —WILLIAM REICHARD,
      "An Open Door" (*This Brightness*, 2007)

A Discourse on Method

If this book is your first encounter with the idea that the mind is a bundle of computations, reading the previous chapter may have been something of a transformative experience for you. Francis Crick, best known for discovering together with James Watson the double-helix structure of the DNA molecule, gave the title *The Astonishing Hypothesis* to a book in which he equated the mind with "the behavior of a vast assembly of nerve cells." How much more astonishing is the hypothesis that what truly matters about the behavior of nerve cells is what they compute!

Whether you have been transformed, astonished, or merely intrigued by this hypothesis (which complements the other one nicely),[1] let me tell you that the best stuff—seeing it grow, prosper, and bear yummily explanatory fruit from one page to the next— is yet to come. So is, I must disclose, some hard thinking, considering what we're up against here. If the mind is a bundle of computations, then understanding how it works would seem to call for a feat of reverse software engineering—a field of endeavor whose defining observation is "Hell is someone else's code."

Fortunately, just as you did not need a degree in computer science to understand the fundamental nature of computation, you do not need one to work out quite an adequate explanation of how computations come together to form a mind. This is because the machinery of mind does not, in fact, use any kind of "software" for anyone to decipher—just as a falling pebble does not use software to plot its course. To start making sense of this machinery, we need to learn to think about the mind on a number of levels.

It is the need to do so that distinguishes mindful entities— those whose representations are used by the entities themselves to some purpose—from mindless ones. It does not take a lot of sophistication for a system to possess rudimentary purposive mindfulness. What are its telltale signs? Imagine yourself waking up in the morning, perhaps a little earlier than usual, to discover that your slippers are slowly but steadily trying to get out of your sight. Such a scene would definitely raise more than one question. Some of the more likely ones are "Why?" (or, if you are prone to taking things personally, "Why are they doing this to me?") and "How?" A little reflection reveals that this last question is too general, which suggests multiple alternatives that are more specific, such as "How do they figure out where to go?" and "How do they actually move?" Now, if you have more than a passing curiosity about your world, you would not settle for having just one of these questions answered. They all address different aspects of the startling mindfulness exhibited by your slippers, and their respective answers tend to complement each other, resulting in a more complete understanding.

For the stealthy slippers scenario, the "why?" question has long been settled by the person in whose mind it originated: Philip K. Dick, the prophetic purveyor of philosophical paranoia, which he dressed up as science fiction and sold to pulp publishers to make a living. In the story "The Short Happy Life of a Brown Oxford," one of Dick's cartoonish scientist characters introduces a "Principle of Sufficient Irritation," according to which inanimate objects, such as shoes, can only take so much abuse before they try to do something about it.[2] Before the shoe's bewildered owner could take up the "how?" question, the newly animate runaway shoe in the story got away in pursuit of a lady shoe love interest, proving the effectiveness of the *cherchez la femme* cliché in settling the "why?" question in the male-dominated genre that was 1950s sci-fi.

The truth, however, is often stranger than even PKD's fiction. Earth's biosphere is replete with microscopic clumps of organized matter—single cells, such as bacteria or brewer's yeast, which we tend to regard as hardly more animate than a shoe—that sense, process, and act on outside chemical information, in the service of higher purposes with which shoe-wearing multicellular hulks can readily identify. One such purpose is sustenance: *E. coli* bacteria, for example, can sense minute directional differences in the concentration of useful metabolites such as aspartate and swim in the direction of the higher concentration. Another purpose is procreation: yeast cells sense and follow pheromone gradients that lead to receptive members of the opposite sex.

The investigation, then, begins with the big "why" and continues with a series of "whats" and "hows." *Why* did a yeast cell cross the road? To get to the block party. *How* could it tell which way the party was? By sensing where the pheromones were drifting from. *What* did it need to compute to find that out? The concentration of pheromones in each direction—the direction in which it was higher points to the source. But if pheromone molecules hit the cell and bounce off it, *how* could it avoid counting the same molecule more than once and getting the direction of the higher concentration wrong? By capturing molecules that hit its specialized pheromone receptors and metabolizing them into some other stuff. (This is the cell's equivalent of counting items off on its fingers and setting them aside.)[3] And *how* did it get closer to its prospective mate? By growing a projection, known as a shmoo (I am not kidding), in the right direction.

Even with unicellular life, there seems to be no end in sight to the march of questions. Entire scientific careers can be (and increasingly are) spent on understanding exclusively some genetic, metabolic, signaling, or structural aspect of life's minutiae, such as growing shmoos in yeast. Interestingly, however, the very same questions apply, level by level, to all purposive information processors, from a single-cell replicator, propelled by a simple mind to seek a mate, to a considerably more complex replicator whose mind, with tragic foresight, "misgives some consequence yet hanging in the stars" and whose pursuit of happiness is soon checked by fortune's hand:

BENVOLIO

Here comes the furious Tybalt back again.

ROMEO

Alive, in triumph! and Mercutio slain!
Away to heaven, respective lenity,
And fire-eyed fury be my conduct now!

Re-enter TYBALT.

Now, Tybalt, take the villain back again,
That late thou gavest me; for Mercutio's soul
Is but a little way above our heads,
Staying for thine to keep him company:
Either thou, or I, or both, must go with him.

TYBALT

Thou, wretched boy, that didst consort him here,
Shalt with him hence.

ROMEO

This shall determine that.

They fight; TYBALT *falls.*

BENVOLIO

Romeo, away, be gone!
The citizens are up, and Tybalt slain.
Stand not amazed: the prince will doom thee death,
If thou art taken: hence, be gone, away!

ROMEO
  O, I am fortune's fool!
BENVOLIO
  Why dost thou stay?

  Exit ROMEO

  Enter CITIZENS, ETC.[4]

Why does Romeo linger at the scene of the sword fight, and why does he eventually leave it? Shakespeare engages us not because his protagonists' motives, decisions, and actions are unfathomable. On the contrary, intuitively we probably know the answers to many of the questions that apply. We should therefore be glad to have come into possession of a conceptual tool kit that is honed so as to dispel whatever mysteries still cling to human nature—those attached to the as-yet-unanswered "why?" and "how?" questions.

Let us arrange our tools on the bench, then. The no. 1 tool is the conceptual lens through which we can see everything as computation. The no. 2 tool is an assay for mindfulness, which helps us separate boring computing for its own sake (falling pebbles or flung penguins) from interesting representational computing (sensing, signaling, and behavior-inducing neurons, cogs, or transistors). The no. 3 tool is a zoom attachment that allows us to focus on various levels of explanation—from the more general "why" questions, whose likely answers will be couched in complex terms that in turn require further explanation (as in "Romeo killed Tybalt in revenge for the death of his kinsman Mercutio"), to the more specific questions about the "how" and the "what" of the corresponding computations (like those carried out by the neuropharmacological circuits that embody what Romeo proclaims to be his "fire-eyed fury" and that are too numerous to list here).[5]

By making use of these tools, we will be able to recognize the *problems* that all minds, natural or artificial, must contend with; to see which *tactics* may help solve the problems; and to convince ourselves that *implementations* of the tactics are possible that rely only on the available means and materials. Some of the necessary work we already did in the previous chapter, when it was made clear that the world is sufficiently well behaved, statistically speaking, to make attempted forethought pay for itself. This is where we should pick up the thread, then, starting with forethought and zooming in and out through the issues as they arise, wherever the quest takes us.

Faster Than a Speeding Marmot

Seeing that forethought is possible in the world we inhabit, how does it actually work? All of us information-processing creatures, from yeast to the citizens of Verona, practice forethought, but the details of how we go about it vary widely.

A lot depends on where you live and who your neighbors are. In many corners of this planet's biosphere, following literally the model of blind Molly from the last chapter (that is, banging your head on the ceiling and digging through dirt in the direction that seems most auspicious, given the echo) would be detrimental to your chances of seeing your kids through college, or at least seeing them have kids of their own—the point at which evolution finally seems to relent from its personal interest in you and you are free to enjoy retirement, if you can afford it.

Because in evolution there is no top of the heap in an absolute sense, it is equally meaningless to claim that the Duke of Verona is either more or less successful than the yeast that froths his beer. It does, however, make sense to ask what general behavioral recipes and abilities are shared by all naturally occurring forethought-practicing entities (and of course what computations are needed to support such abilities). There is little integrative work on this question in cognitive science (neuroethologists and neurobiologists tend to study particular behaviors or subsystems in certain species, and roboticists too shun generalities, all for very understandable reasons),[6] and so common sense will have to do.

Take, for example, this superpower that I possess: dodging marmots. It may not sound like much, but unlike the superpowers of comic-book heroes and some misguided citizens,[7] mine is real, as I was able to verify a few years ago during a hike in Mt. Rainier National Park, which is near Seattle. A self-propelled marmot is very easy for a primate like me to dodge. I can sense its approach while it is still at a safe distance, even if it is trying to sneak up on me; I do it by intercepting and measuring the ambient light that is reflected from it and process the images to recognize it as a marmot and to estimate its distance, heading, and speed. I can use my intuitive knowledge of physics (distilled from years of experience going to and fro on the earth and walking up and down on it) to extrapolate the marmot's trajectory and to decide whether or not I need to act. If I do, I send a series of signals to my muscles that bring about the intended end result, which is a frustrated marmot. Voilà!

It would be a more serious matter if marmots were ever used against me as missiles by an ill-wisher. If it was hurled by hand or dropped from a balcony, I might still be able to see the incoming marmot in time to step aside, at least if I acted sharp, whereas a drone-launched marmot could easily prove lethal. The worst scenario, however, would be one in which I, for whatever reason, no longer cared. A depressed person, who lacks the basic motivation to act, even in self-defense, is quite easy to hit with a marmot.

This last observation explains why the general recipe for managing behavior through forethought must begin with the need to get off one's behind.[8] Once you do, you can sense things before they happen to you, plot a course of action based on the sensed data and on past experience, and execute the chosen maneuvers. The ingredients of this recipe are familiar to every psychology student under their textbook labels: motivation, perception, thinking, and motor control.[9] They are easiest to understand when taken up in a somewhat different order. It begins with perception, a clear view of which is critical for grasping both how possibilities for action present themselves to minds and what motivation means computationally. Because thinking is where the main potential for complexity and sophistication in cognition is found, most of the thinking about thinking will be postponed until later in the book.

A Treatise of Human Nature

Even if it seems natural to conceive of perception as a window through which the brain, in the safe house of the skull, monitors the neighborhood for any developments or portents, this conception is wrong. Suppose that an embrained but as yet sightless mind, desperately in need of distraction from its confinement, decides to watch a football game. Sawing through the front of its body's face to make a window to let the light shine on the brain would be as futile in trying to get it to see as installing a little TV screen on the inside and connecting it to a camera looking out. To see things through such a window or on the internal screen, the brain would need something like an eye, along with the machinery that processes what the eye tells it. This, of course, is exactly what a regular human brain already has, which demonstrates that the window/TV simile merely postpones the explanation instead of providing one.

The path to a real explanation begins with the realization that vision and other senses deliver variously *interpreted* representations of the world rather than snapshots or recordings of it. Thus, it is more useful to think of the process of seeing a football game play out not as a video transmission for the benefit of an internal viewer, but rather as a radio show in which a boxful of commentators, speaking all at once, describe the action to a bunch of listeners.

Why multiple "commentators"? Because an analogy with a single running commentary would only ever suffice to explain a very, very simple perceiver, such as perhaps a household thermostat, whose comments on any action in the house would go like this: " . . . ; too cold; too cold; just right; just right; just right; too warm; just right. . . . " A slightly fancier thermostat that also includes a temperature display provides two related but not identical commentaries on the same situation: one is the on/off track, delivered exclusively for the benefit of the air conditioner, and the other is the number in the little window that shows the temperature. An even better one could include a humidity reading, which perhaps could be wired to a humidity control device that is separate from the air-conditioning unit.

The possibility of a perceptual information stream having more than one destination explains why in drawing an analogy between perception and a radio show I made a point of including not just multiple "commentators" but also multiple "listeners." At the very least, the household climate control system just described must have *two* distinct "listeners" that use the information provided by the sensing channels (the hygrometer and the thermometer): one in charge of the humidifier and the other in charge of the air conditioning. So, if anyone attempts to sell you the idea of a single, indivisible black box labeled THINKING as an explanation for how perceptions give rise to actions, expose it as sham by asking, "Yes, but what's going on inside the box?"

As this example shows, because even minimally complex minds simultaneously track multiple aspects of the environment while at the same time controlling multiple means of acting back, their innards must be *distributed*. This means that minds are composed of at least several—and possibly very many—interacting but distinct functional parts. (Unlike a physical part, such as the wheel of a car, a functional part is a role that can be played by different physical parts. For example, many kinds of vehicles have the functional part "support," which in a car is played by wheels, in a tank by treads, and in a sled by runners.) The functional parts, their relationships (which may be hierarchical, as when some parts are composed of several others), and the interconnections and interactions among themselves and with the outside world together determine the kind of mind that arises from all this bustle.

Because what matters about a mind is its functional organization (and not, as we learned at the end of Chapter 2, the stuff it is made of), sharpening our thinking about functional analogies can really help us understand how minds work. It would be particularly helpful to come up with some down-to-earth example of distributed organization, seeing how essential it is to the architecture of minds. We know that on the perception (input) side there are many sensors that send their signals in all at the same time. (Each light-sensitive cell in your retina is one such sensor.)

We also know that on the action (output) side the situation is similar: each of your muscles consists of many bundles of individually innervated fibers, and each joint is served by many muscles. There is no reason to assume that in between perception and action—in processing, or thinking—things are any different (more about this in later chapters).

Is there a good functional analogy, along the lines of the radio show example, that covers all these bases? Yes, there is: a parliamentary democracy.

The roots of this analogy go back to the ideas of the great Scottish Enlightenment philosopher David Hume. In a book titled *A Treatise of Human Nature*, published in 1740, Hume wrote:

> I cannot compare the soul more properly to any thing than to a republic or common-wealth, in which the several members are united by the reciprocal ties of government and subordination. . . . And as the same individual republic may not only change its members, but also its laws and constitutions; in like manner the same person may vary his character and disposition, as well as his impressions and ideas, without losing his identity. Whatever changes he endures, his several parts are still connected by the relation of causation.[10]

Hume's analogy, with which he clarified his revolutionary (and, as we shall see, prescient) stance on personal identity and the nature of the self, is directly relevant to the theme of this book, if only because it implies that a person's pursuit of happiness is best viewed as a kind of mass marathon of the mind's multiple constituents, not a solitary trek of an indivisible ego. In reality, the republic analogy applies all across mind science. To see why this must be so, we need only recall that minds are made of computation and that what Hume called "the relation of causation"—the causal organization of a system, whether political or cognitive—is what defines both computation as such and the use of computation by minds through the mechanism of representation.

Because the architecture of each of the mind's top-level functional "parts"—perception, thinking, action, and motivation—is distributed, the republic analogy applies not just to the mind in general but also within each of those domains, at level after level. (A computer scientist would say that it applies recursively.) A parliamentary democracy founded on the principle of separation of powers governs itself by balancing legislative, executive, and judiciary activities, which makes it functionally distributed at the top level. Each of the branches is, in turn, distributed. Interestingly and importantly, this includes the executive power, which in present-day Britain, for example, is vested in the Cabinet of Her Majesty's Government (over which the titular monarch, thankfully, has absolutely no control). Hume's point is that even under a regime in which certain executive decisions are made by a singular legal entity (as in the still largely autocratic Britain of Hume's own time or in the present-day United States), they are in fact made by a plural cognitive entity—the republic of soul.

Perception by Numbers

If a person's mind (of which his or her soul is, as we shall see later, a proper part) is like a democratically governed commonwealth, then perception is the array of information sources, from mass communication media to targeted intelligence-gathering, that the cabinet members use in formulating the foreign policy.

Because for the cabinet perceptual input is by definition the sole origin of information about the outside world, we should really think about it as meeting in an underground bunker. This important detail adds an interesting twist to the idea of the mind as a democracy. As it turns out, a narrower analogy works even better: the mind is really like a wartime democracy (think World War II Britain).[11]

This line of reasoning shows just how important perception is for the functioning of a human mind. The availability of reliable information about the outside world does not guarantee sane "foreign policy" or effective conduct of the metaphorical war for survival on the part of a mind. (Human societies too, republics or not, are inordinately prone to suffer themselves to be governed by those whom the prophet Jeremiah described as "foolish people, and without understanding; which have eyes, and see not; which have ears, and hear not."[12]) However, the *absence* of perceptual information definitely complicates flight from peril and effectively dooms any attempted pursuit of a mate, let alone of abstract happiness. More than that, seeing that the contents of a human mind do not get downloaded into it fully formed, we realize that the cabinet members in our analogy must have been inside their bunker *all along*. This insight suggests that perception is indispensable not only for guiding immediate "here and now" behavior but also for driving and sustaining *development*—the protracted process that transforms a bunker-bound crèche into a war cabinet.

The array of primary information sources in human perception is literally an array—of numbers that stream into the brain from an assortment of measurement devices. The sense of sight begins with hundreds of millions of measurements of electromagnetic energy, carried out by the photoreceptors that absorb the energy of light focused by the eye's optics onto the retina; each eye ends up sending on to the brain about one million fibers that carry an already heavily processed array of visual information. Hearing originates with tens of thousands of inner-ear hair cells, which transduce the mechanical energy of sound into neural firing. The sense of touch is mechanical too, with receptors scattered throughout the skin and the mouth. Then there are two chemical senses, taste and smell, whose receptors measure the concentrations of thousands of types of molecules of interest in their vicinity. Finally, there is the sixth sense, interoception—a motley collection of internal mechanical, chemical, and thermal gauges that report the body's vital signs to the central nervous system.

To recognize the vastness of the computational problem faced by any mind that is bent on seeing, whether it looks at the world through Romeo's eye or a robot's megapixel-resolution camera, consider this: it must deal with a torrent of data that delivers several times per second a new $1,000 \times 1,000$ table of numbers to be made sense of. That's all there is: a constantly changing array of numbers in which many things, some potentially interesting or dangerous, are lurking—Juliet on her bedroom balcony, the flowerpot next to her on the parapet, Benvolio's face, Tybalt's sword.

To salvage from the data deluge some useful information, the mind's only recourse is to try to *relate* some of those numbers to others.[13] For starters, it would be nice to be able to do something to ensure that the image that falls on the sensor is sharp. Whereas even a blurred image of Juliet would not look anything like a flowerpot (or so one imagines), the possibility of a momentarily myopic Romeo mistakenly skewering Benvolio instead of Tybalt seems quite plausible.[14]

How should numbers arranged in a table be compared to one another so as to reveal whether or not the image they form is focused? It's actually quite easy (which is why every modern camera has built-in autofocus). The idea is to scan the table while comparing the values of adjacent entries. Scanning a blurry image generally yields a series of numbers that change relatively slowly (and under an extreme blur, as in dense fog, not at all). In contrast, in a focused image the transitions between adjacent numbers are every now and then quite sharp. (For instance, if you sit across from me, this happens in each place where it appears to you that my face ends and the wall behind me begins.)

What the control mechanism (brain or camera) needs to do, then, is to keep changing the lens focus little by little while computing those local-neighborhood sharpness estimates, until their outcome is satisfactory. For this simple trick to work, the neighboring numbers must correspond to—represent—neighboring directions in the visual world, which they indeed do. Physical law ensures that the lens of an eye preserves visual neighborhood structure in the image that it projects onto the retina, and evolutionary pressure has already seen to it that the retina preserves this structure while converting the image into an array of numbers.

Being able to focus on a scene is a far cry from being able to interpret it: my digital SLR camera focuses like a fiend, but understands nothing. I love my old SLR and will not trade it for a smarter model, but for someone who values smarts over sheer versatility and obedience, the temptation is growing apace with technology. For example, younger-generation cameras these days are getting pretty good at telling apart faces from other objects, which allows them to focus automatically on people if their master cannot be bothered with focusing by hand. Still, it will be some time before cameras get to be as good as people are at recognizing people. How do we do it?

As the no. 3 conceptual zoom tool from a few pages back kicks into action, we discover that, as always in cognition, the big "how?" question spawns a whole spate of smaller ones, some still very general, others quite specific. How is it possible to tell whether or not the object that is being looked at is a face? How can one determine whether or not a given face is familiar? And if the face is familiar, how can its identity be established? A complete answer to any of these questions would consist of two parts: one that describes the computation ("Take the array of numbers that represent the scene and carry out the following computations [detailed step-by-step instructions omitted]"), and another that describes how the computation is carried out by neurons ("The axons that form the optic nerve connect to neurons in the lateral geniculate nucleus of the thalamus according to the following pattern [detailed wiring diagram and neural activity reports omitted]").

Although not all the details that I boldly glossed over just now are known, cognitive science has a pretty good "big picture" of how face recognition works.[15] It is easiest to start with the last and most specific of the "how?" questions posed above, the one concerning face identity. By definition, a visual system can only recognize ("re-cognize") a face as belonging to a particular person if that person had been seen at least once before. It would seem that recognition, then, is a simple matter of memory storage: just save a "snapshot" (array of numbers) for each face you see and later compare the representation of a face that needs to be recognized to each of the stored snapshots.

The main problem with this idea is that the same face can look very different—that is, it can present a very different array of numbers—depending on how it is illuminated and oriented with respect to the viewer. This is why at a campfire event you can reliably scare a five-year-old by illuminating your face with a flashlight from below. (Warning: failure to revert the illumination direction to normal promptly enough may result in everybody's evening being ruined.) This is also why sheep, who can be quite good at recognizing each other by face, are baffled by upside-down pictures of their acquaintances but monkeys are not. (One expects that a circus troupe of sheep trained to perform on the trapeze would be more tolerant of face inversion.)[16]

Even though faces (or any other visual objects) cannot be reliably recognized through exhaustive number-by-number comparison to stored snapshots, a mathematical analysis of the recognition problem suggests a modification to the store-and-compare procedure that makes it work.[17] The idea is really very simple: make the comparison approximate. A slavishly literal comparison between two arrays of numbers is all-or-none: a difference in even just one place brands the snapshots as "different." In contrast, if the comparison procedure is made to estimate the *degree* of difference between the snapshots, its outcome is a graded quantity that reveals how (dis)similar they are from each other, instead of merely stating that they are not the same.

To appreciate the informativeness of representations that are based on graded similarity, imagine having witnessed a crime and being asked to identify the suspect in a police mug-shot album. Under a strict match regime, you'd probably have to rule out all the candidate mugs because none of them would coincide exactly with the face you remember. Not so under graded comparison: by pointing to several mugs that resemble to various degrees the person you saw, you could narrow down the range of possibilities, thereby discharging your civic duty and making your city's streets safer.[18] Even better, graded similarity–based representations are not only informative but also frugal in their demands for memory. Once you have stored a sparse "starter" set of faces in a relatively detailed snapshot form, each new face can be represented by just a handful of numbers that stand for its similarities to the stored snapshots.

Representation by similarity works on the same principle as a global positioning system (GPS), an electronic navigation aid that so conveniently absolves us from the need to know how to get where we're going. A GPS receiver figures out where it is by estimating how far it is from each of the orbiting satellites, whose own positions are precisely determined at all times. Likewise, a cell phone can triangulate its location from the strengths of the signals it receives from several relay towers. This robust analogy between positioning by triangulation, on the one hand, and representation by similarity, on the other hand, leads us right up to a key conceptual tool for understanding how the mind works: representation spaces.

Representation Space: The Final Frontier

To stake a claim in the conceptual frontier-land of mind science, think of each face snapshot as a point in a big representation space—let's call it the *face space*. (The concept of a point in a representation space is so widely applicable that we might just as well give it a name: this is our no. 4 conceptual tool.) Storing a bunch of face snapshots then becomes a matter of marking their representations in the face space; think of sticking little flags on a map in the war room. Once the "flags" are up, a new face can be described precisely and efficiently by how close it is to each of these reference points, because proximity between points in face space corresponds to similarity between the faces.

If a face-space point represents a particular snapshot, then all the different directions away from it correspond to all the different ways in which the appearance of the face can change. We already know that some of these changes actually need to be ignored because they stem from illumination or orientation shifts and are of no consequence to face identity. A smart procedure for gauging face similarity—that is, face-space proximity—must therefore give less weight to these face-space directions and more to those that correspond to identity differences. It turns out that brains can learn this kind of smarts on the job, simply by storing a few examples that represent each kind of change.

A stored example is represented in the brain by the standard building block—a neuron. A typical neuron in the visual system can learn to represent a snapshot of a face (or of some other object) by being "imprinted" with it—having its input synapses adjusted so as to evoke a selective response to the stimulus in question. Such learning results in the neuron becoming tuned to the stimulus, so that subsequently it responds the strongest to it and progressively less strong to stimuli that are less and less similar. An ensemble of such coarsely tuned neurons has exactly what is needed to pinpoint the face-space location of a new face, using triangulation by graded similarity (the same computation used by GPS receivers).[19] But how can this mechanism distinguish between important and non-important face-space directions? By seeking regularities in the skein of face-space tracks laid down in the representation space by individual faces as they are observed under varying conditions.

As Romeo sees Juliet for the first time, some of his neurons that usually respond to visual objects become active, each one just so—depending on how excited it gets by being exposed to Juliet's features, which in turn depends on how close these are to this particular neuron's prior experience. The list of numbers that denote the activities of this ensemble of neurons defines the point in Romeo's face space that lights up in response to Juliet's face (as Shakespeare almost wrote—"It is the brain, and Juliet is the sun"[20]). Here's what happens when Juliet turns aside: as the orientation of her face changes gradually, its representation in Romeo's face space splits off from the original point and gradually moves away, along a very specific face-space track. Romeo has never seen Juliet's face from the side before; will his visual system now do the smart thing?

To realize that Juliet seen from any angle is still Juliet, Romeo's visual system can draw on the memories of its experience with other faces. By the time he meets Juliet, he has seen other faces undergo the same turning-aside transformation. For each of those faces, he has stored a sparse sequence of representations that blaze the "turning-aside" track across face space. He can now use this knowledge, distilled from the regularities observed in the past experience, to interpret the ongoing change in the appearance of Juliet's face.

Distilling knowledge from visual experience puts available memory to good use. Instead of grabbing face snapshots indiscriminately, the visual system samples and strings them together in strands that run through the face space in parallel to each other, corresponding to different faces undergoing the same transformation. As a result, it can make up for changes in the appearance of a new face through a kind of analogy, by interpolating its prior experience with other, similar faces. The very same trick also works for other categories of visual objects and, interestingly enough, for other cognitive tasks, such as motor control, reasoning, and language: knowledge grows out of experience, and analogy rules all.[21]

Being in the World

For the mind's war cabinet that meets in the dark shelter of the skull, the patterns of data that flicker across the sensor arrays are the only signs that it can ever have of the external world. What if there is danger out there? Imagine how terrifying it would be to know that there is a madman with a sword a few paces away from you, but not whether he is before or behind you. For obvious evolutionary reasons, such an ability to recognize the "what" without the "where" of a stimulus is unheard of in animal cognition, where even a yeast cell's chemical sense is directional.

In us primates, who are equipped with camera-like eyes capable of highly directional quality imaging, the face space and the representation spaces for other objects are all yoked, as it were, to a common "space" space. This common underlying spatial scaffolding supports not only vision but also other senses, the data from all of which are brought into spatial register. This sensory integration effectively creates a complete virtual world, centered on the perceiver.

The realization that one's perceptual world is a virtual construct (even if it does reflect faithfully many aspects of external reality) may take some time to sink in. As a first aid in overcoming the understandable, yet false, belief that minds have an unmediated grasp of the real world, ponder this: your vantage point in this world seems to be located right behind the bridge of your nose, yet what you see in front of you is this page, not the inside of your skull, where, as I keep pointing out, it is perpetually dark. This little observation suggests that our perception of the external world, effective as it may be in delivering those cues that matter for our evolutionary fitness, is not to be trusted blindly in all matters. And yet, trust it blindly we must, because we have no choice: the cabinet members sequestered in the war room may become collectively aware of their predicament, but they may not escape it.[22]

The predicament of being confined to a virtual reality rig may feel disturbing when one first becomes aware of it, but in practical terms it is quite benign, because the rig has evolved to function well enough in the environment it is situated in. The first order of business for the senses is to make the spatial structure of this environment available for the decision-making processes in the rig. Given that sensations are spatially tagged, the proper way of describing how they present themselves for the executive cabinet's consideration is to envisage a grand annotated map that wraps around the war room.

The grand map represents the surrounding space, which is how it conveys the "where" information. It is also annotated with "what" information in the form of multiple face/object spaces, each attached to some location on the master map. The function of this map is to generate behavior by coordinating perception and action—battle intelligence and battle plans. It does so by expressing the inputs from the senses and the potential outputs from the behavioral control processes in the same embodied and situated language of cues and action "handles," all anchored in extra-personal space.

It is this kind of battle intelligence that guides young Romeo's behavior as he scales the wall of the Capulets' orchard, on the night of the masked ball, where he is about to fall for the daughter of the lady of the house. A pale visage appears in the dusk. On the grand map in Romeo's brain, it receives a tentative label of "female face" and is represented by several numbers conveying its similarities to the faces of some women known to him. It is attached to a particular location in the visual field: up there, on some kind of balcony. He is drawn to it; he approaches. The face, as it comes into focus, resolves itself into that of Juliet.

The Instruments of Change

The grand map that presents itself to the mind's executive cabinet is far removed from the raw data gathered by the senses. It is loaded with actionable, useful information, which the perceptual processes work hard to extract and refine. What form does that information take? Sci-fi movies that involve robots occasionally offer the viewer a glimpse of an imagined cybernetic protagonist's display-like internal map, on which various objects of interest are labeled in plain English (set in a typeface intended to look futuristic). This cinematic trope does get one thing right: any embodied agent, including us biological robots, needs to coordinate perception and action, and a map is a very convenient prop for doing so. There is also, however, something very wrong here: a map that is annotated in English cannot be a *part* of a mind, because it would only be intelligible to a *whole* mind that can read. The labels on the human mind's grand map are not human-readable, because the consumers of the information that they carry are not human: they are the many computations that may in themselves be simple, yet are collectively complex enough to be a mind.[23]

Some of these computations are directly aimed at making things happen by steering the body to physically engage the rest of the world. A mind can only fulfill its function of channeling forethought if it is capable of bringing about change, by moving the body of which it is part, or by using the body to move other things. As in perception, in the control of movement there is a need for representations that downplay irrelevant variation, such as the postural context in which a reaching movement needs to be executed. This is why motor control "scripts," like perceptual representations, rely on similarity to stored examples and, more generally, on analogy. You learn to balance yourself on a snowboard by just doing it (on moderate enough slopes) while retaining and organizing motor memories of the more successful of your moves; the resulting experience is likely to help you somewhat in your first attempts at surfing, despite the differences between the kinds of support and resistance offered by snow and water.

Animal brains control the posture and movement of the bodies in which they reside by sending to muscles signals that cause them to contract and exert force. To move a body around, or even just to prevent it from collapsing on itself like a skin stuffed with organs, a brain needs to compute some numbers, one per muscle, and send them to their destinations. Some simple actions may be completely specified by a single number, as in the case of a scallop closing its shell in response to the passing shadow of a cuttlefish. In comparison, to animate a human skeleton, a bundle of scripts controlling a multitude of muscles and unfolding in lockstep need to be played out, each consisting of a sequence of numbers, generated in the proper order and with proper timing.

These numbers, their order, and their timing all depend very much on the mechanics of the body part or parts that the script needs to control and on the kind of environment in which the body is situated. In a body that has a few hundred muscles with which to pull itself around, the computational problem of motor control is difficult indeed. Which muscles to activate, how strongly, for how long, and in what order—all these details need to be figured out, and this needs to happen in a timely fashion (whether to avoid becoming someone's dinner or to turn someone else into your dinner, bar access to your gene pool, or gain access to someone else's). Because of all that, in behaviorally sophisticated animals with mechanically complex bodies, such as humans or ravens or octopuses, motor control has to be hierarchical, with simpler muscle synergies serving as building blocks for the construction of progressively more and more complex ones.

The motor control problem can be made more tractable by perceiving in the geometry and physics of the body and the environment not just obstacles that must be overcome but opportunities that can be exploited. Such opportunities are called *affordances*: a flat horizontal surface affords sitting on, a basketball backboard affords directing the ball into the hoop, and a pond surface affords running on if you are a water spider or a basilisk lizard.[24] Experienced biological perceptual systems seek out affordances automatically, because in animals perception and motor control are inextricably interwoven: a body can get eaten because of a failure of either.

If predators, prey, competitors, or other objects of interest in your ecological niche move fast, and if your performance depends on matching their speed, your motor control system cannot rely too much on incremental corrections driven by perceptual feedback. The only resort in such cases is to distill behavioral experiences into task-specific models that capture as much as possible of the mechanics of your own body ("if *these* commands are sent to *those* muscles while I am in *that* posture, I duck"), the situation ("if a sword thrust is coming from *there*, duck"), and, if the situation involves other animate agents, their own statistically likely behavior ("what would Tybalt do?"). Such process models are embodiments of forethought, used to *simulate* ahead of time both what may happen in a highly dynamic situation and how to deal with it, in the most literal physical sense.[25]

Internal dynamical models of the world are the richest and most intricate of the many sources of information that interact with the mind's grand map and help it shape the battle plan. Because they extend into the future—or rather, into the many simulated possible futures that build on past experience and interact with present actions—the grand map describes, over and above the concrete spatial disposition of the immediate environment, an abstract space of possible actions and consequences. Computationally, the influence that the mind exerts over behavior thus reduces to navigating a labyrinthine web of cause and effect. Imagine yourself standing at a crossroads in this maze; where would you go next?

The Value of Everything

The executive function of the mind is often referred to as the closing of the perception-action loop, which makes it sound as if there is a single perceptual item on the agenda that inexorably leads to a single action. In reality, the task of connecting perception to action via the grand map has all the characteristics of a holy mess. As should be obvious by now, the work of the mind's war cabinet is never complete: far from being static, the grand map is constantly updated. The dynamics of the update cycle reflects external events (as when danger looms or an object of some other kind of potential interest comes into view), as well as the mind's own internal needs. All these processes flood the map with content, which cannot all be dealt with at the same time. Getting the virtual ducks in a row is therefore a necessity, and so the grand map includes, in addition to information about objects, their locations relative to one's vantage point, their expected behavior, and the potential actions that can be directed at them, estimates of their *value*.

Although the relative value of various goals or courses of action can be pondered (and often *is* pondered, ad nauseam, as in the case of a certain prince of Denmark whose native hue of resolution was sicklied o'er with a pale cast of thought), in natural cognition valuation does not always require deliberation. Indeed, value is a common by-product of experience, including experience that, over evolutionary time, helps shape the animal's genome. Being a distillate of experience, value is fundamentally a matter of statistics (which, as noted before, does not mean that valuation is never certain, but rather that value-based decisions deal with the ubiquitous uncertainty of life using the computationally appropriate means). A handy example of the statistical underpinnings of value is found in the pattern of judgments people make of faces—by far the most important class of perceptual objects for social primates.

Prompted by the common saying that beauty is in the eye of the beholder, experimental psychologists had human subjects rate the attractiveness (which is a kind of value) of composite faces that had been generated by a computer program by averaging carefully aligned photographs of real people. These studies found a strikingly strong correlation between the perceived degree of attractiveness of a composite face and the number of individual photographs that were averaged to produce it. The beauty of a face, as it were, proved to depend on its location in the subject's face space: the closer a given face is to the center of the statistical distribution, the more attractive it seems.[26]

Insofar as the statistics of experience vary between individuals, beauty is subjective. A face that causes my heart to skip a beat may appear to you as plain—or worse, if you and I are of different species. (This knowledge almost manages to spoil for me the scene in Tolkien's *Lord of the Rings* in which Gimli, son of Glóin, declares Queen Galadriel to be the "fairest"; a Dwarf could never see an Elf as anything but hideously ugly.) For stimuli other than faces, certain characteristics of perceived attractiveness, or value, do seem to be widely shared, perhaps because they are determined by computational factors that are common to all sentient beings. For instance, the few existing studies of landscape valuation by human subjects suggest that complexity, openness, and water features contribute to scenic beauty and attractiveness.[27]

Although no comprehensive computational theory of scenic beauty exists as yet, my guess is that a major determinant of such beauty will turn out to be novelty, along with a perceived promise of novelty, such as offered by a new vista and a mountainous landscape, where every turn brings new tidings. Even hobbits, renowned for their deep love for the modest and cozy Shire, fall for grand vistas when they see them; hence Bilbo's self-confessed longing for Rivendell: "I want to see mountains again, Gandalf— mountains."[28] In *The Water of the Wondrous Isles*, William Morris, from whose work Tolkien drew much inspiration, imagined a young woman's first glimpse of a mountain range thus:

> At last the trees failed them suddenly, and they came forth on to a wide green plain, all unbuilded, so far as their eyes could see, and beyond it the ridges of the hills and blue mountains rising high beyond them. . . . When Birdalone's eyes beheld this new thing, of a sudden all care left her, and she dropped her rein, and smote her palms together, and cried out: Oh! but thou art beautiful, O earth thou art beautiful![29]

Things Get Interesting

The appearance on the grand map of the mind of value, over and above mere information, is literally what makes life interesting, both for better *and* for worse. Merely singling out an object of desire automatically devalues other objects, just as perceiving something as worthy of avoidance makes the alternatives more appealing. To withstand evolutionary selection pressures, valuation processes must draw upon observables: it would be stupid (and in the long run suicidal) to attach to an object or a course of action a value that does not in the least depend on it. This means that valuation is not an afterthought or an add-on but an integral part of natural perception: all that you see, all that you hear, all that you taste, touch, smell, and feel is weighed for its potential value to you.[30]

To be of use in guiding behavior, the values carried by all the perceived objects in the environment and by every action that it affords must be expressed in a common currency, so that they can be readily compared. It turns out that our brains do indeed maintain representations in which chunks of information and fragments of action plans are brought to a common denominator and used to guide behavior on a moment-by-moment basis. For obvious reasons, the emerging discipline that studies this process and the representations it deals in is called neuroeconomics.[31]

We saw how natural it is to think of the representations used by perceptual processes as spaces (as in the face space theory). We now know that motor actions, being represented as patterns of numbers over which similarity relations hold, admit the same formalism: just as a face can be thought of as a point in a face space (spanned by some stored exemplars), so an action is a point in an action space, which likewise arises from experience. Now join perception and action together and throw in value; what you get is an abstract "value terrain" in which locations stand for perception-action bundles and elevation represents attraction—the deeper and steeper the attractor valley, the more powerful its draw.

This insight implies that the mind's grand map, far from being flat, is shaped by external affordances and by internal desires and fears into a convoluted landscape that steers behavior like the bumps and pegs in a pinball machine. The buckling of the value landscape that is no longer indifferently flat into domains of repulsion (hills) and attraction (valleys) sets in motion the pursuit of happiness.

The notion of happiness implicit in this observation is radically minimalistic. For happiness or misery to emerge, indifference must be broken, and so the pursuit of one and, symmetrically, the avoidance of the other can be regarded as the ultimate source of motivation. No matter what else can be said about happiness, whenever you find yourself motivated (that is, being moved) to do one thing rather than another—a predicament that incorporates what we perceive as choice—happiness is the prime mover.

This incipient computational understanding of motivation brings with it a measure of insight into the stuff that happiness is made of and reveals why it is so notoriously fickle. The value terrain whose shape steers behavior is itself constantly reshaped by a multitude of factors, not the least of which is novelty. Let us return to Shakespeare's Verona, to the Capulets' ball. Romeo, being a Montague, is not invited, but he sneaks in nonetheless, prodded by Benvolio (whose mocking of the lad's sincere, albeit mercurial, ardor suggests that Shakespeare was being somewhat sarcastic in naming this particular character):

BENVOLIO
>At this same ancient feast of Capulet's
>Sups the fair Rosaline whom thou so lovest,
>With all the admired beauties of Verona:
>Go thither; and, with unattainted eye,
>Compare her face with some that I shall show,
>And I will make thee think thy swan a crow.

ROMEO
> When the devout religion of mine eye
> Maintains such falsehood, then turn tears to fires;
> And these, who, often drown'd, could never die,
> Transparent heretics, be burnt for liars!
> One fairer than my love? The all-seeing sun
> Ne'er saw her match since first the world begun.[32]

It is Romeo's appearance in disguise at the Capulets' feast, where he first sees Juliet and where Tybalt overhears him admiring her, that ultimately brings tragedy to the two great houses of Verona.

It may be crass of me to make the point I am about to make right after quoting from a story of such woe, but I would like, nevertheless, to draw your attention to the bright side of human nature hinted at by this example, to wit: if you are after happiness, change is good, even if, for the moment, it is change to the worse—unless, of course, one is trapped inside a Shakespearean tragedy in which everybody is doomed to die before their luck turns to the better. As Tennyson asserts in *Ulysses*, "death closes all."[33]

As we get older, in our bones we feel less for young Romeo and more for Tennyson's aged Ulysses, who, having striven with gods and brought doom upon Troy, is presently "made weak by time and fate." This increasingly insistent intimation of personal finitude—surely a human physiological universal—is neatly balanced by a belief in life after death in a state of eternal bliss, a concept that is a cultural universal, inasmuch as it is perfectly intelligible even to people who do not succumb to it.[34]

Among the various grounds on which the members of the masses may resist this particular bit of opium, there is the appreciation of how difficult it is for a human being to experience sustained happiness for any prolonged period of time. It seems that there are quite a few folks around who, if the price of bliss is boredom, would choose good entertainment over permanent happiness. Perhaps in recognition of this little problem with the naive promise of Paradise, the path to personal salvation preached by the Buddha has the avoidance of *un*happiness as the key motive, the Awakened One having made it very clear that he did not deem the frequent downsides of life worth suffering for the sake of the occasional up.

The final destination of the pilgrim who sets out on the Buddha's Eightfold Path is a state of liberation that is nowhere nearly as easy to understand as a simple readmission into Eden. On some accounts, the state of nirvana implies cessation of cognition as we know it. To cease having desires, desirable as it may be in view of this doctrine's calculus of worldly suffering, means to cease being human.[35] Moreover, the journey to this destination is irreversible, because once it has been attained, the desire cannot possibly arise in the pilgrim to reconsider and return. On other accounts, however, those who walk this path become more, not less, human—not by rejecting their nature, but rather by gaining insight into it and thereby learning to live the way we ought to live, given what we are.[36]

In the hope of being able to do just that, it makes sense for us to try to learn some more of what cognitive science can tell us about the computational dynamics of the mind and of happiness. An interim conclusion that goes only a little way beyond what we have already learned in this book, yet that can serve as a bridge to its remaining chapters, suggests itself:

> The goal is only a means. . . . Happiness is not in happiness itself, but in running toward happiness.
>
> —ARKADY AND BORIS STRUGATSKY,
> *Noon: XXII Century* (1964, p. 283)[37]

❧ SYNOPSIS _____

The proper way to explain the building blocks of the mind, starting with perception and ending with action, is to consider them as a web of interlinked computations. Although these computations can be very complex (some still elude complete understanding), many are based on simple and intuitive principles, which can be readily illustrated on examples drawn from familiar everyday situations.

In the case of perception, the data that the brain computes with arise from measurements that the senses—vision, touch, hearing, and so on—perform on the environment. The goal of perceptual computation is to recover from those numbers useful representations of stuff that's happening "out there" in the world, out of the direct reach of the brain's neurons. For instance, although the visible shape of a sleeping cat changes as it wakes up, uncurls, stretches, and walks away, the cat is never perceived as deforming, because your visual system factors just the right kind of shape variability into the computations it performs in order to arrive at a stable representation of cathood.

Acting on your perceptions of the world likewise amounts to computation—in this case, computing the signals that need to be sent to your muscles so that their joint action makes your body do the right thing. But what is the right thing to do in a given situation? That obviously depends not just on how you perceive the world or how you can act on it, but also on your values, goals, and motives. All of these aspects of you are at bottom just dynamic arrays of numbers expressed by activities of various cliques of neurons, as are perceptual states and possible actions. Their respective values get translated into a common "neuroeconomic" currency, giving rise to a grand map that shapes behavior by making some courses of action more attractive than others.

Computationally, therefore, the unfolding of behavior can be thought of as a ball rolling down a continually shifting landscape of possibilities, always seeking the deepest valleys. This computational understanding of the nature of perception, motivation, and action offers some intriguing insights into the meaning of, and the prospects for, the pursuit of happiness.

4 | Learning to Think for Yourself

Ulysses ascendant. Remembrance of things past and future. Why everything important that you know you must have learned for yourself. Mirroring the world, mustache and all, one step at a time. Where was I? A moveable feast.

He had never dwelled on memory's delights. Impressions slid over him, vivid but ephemeral. A potter's vermilion; the heavens laden with stars that were also gods; the moon, from which a lion had fallen; the slick feel of marble beneath slow sensitive fingertips; the taste of wild boar meat, eagerly torn by his white teeth; a Phoenician word; the black shadow a lance casts on yellow sand; the nearness of the sea or of a woman; a heavy wine, its roughness cut by honey— these could fill his soul completely.

—JORGE LUIS BORGES, *The Maker* (1960)

It's a poor sort of memory that only works backward.

—LEWIS CARROLL, *Alice's Adventures in Wonderland* (1865, ch. 5)

Ulysses Ascendant

The greatest hero of *The Iliad*, mighty Achilles, was not a happy man. One does not exactly expect to hear much about happiness in an epic in which the earth before the walls of Troy periodically "runs red with blood" and in which Scamander, the god of a nearby river whose course Achilles stops with bodies of dead Trojans, pleads with the hero to do his "grim work" on land.[1] Still, even against that bloody backdrop, Achilles stands out as a particularly dour fellow, whose repertoire of moods is limited to sulking, brooding, mourning, and being beside himself with rage. There is exactly one occasion on which Homer calls him "glad": when his mother, the goddess Thetis, presents him with armor wrought for him by Hephaestos, the blacksmith and artificer to the Olympian gods.

The most remarkable part of that armor, *The Iliad* tells us, was the great shield, which Hephaestos covered with images of the earth, the heavens, and everything in between—the moon, the sun, and the stars; cities (one at peace and one at war with its neighbors), fields, vineyards, pastures, animals, and people. Curiously, the world carved on the shield did not appear to be frozen in time: night and day followed each other, and animals and people went about their business, cattle and sheep grazing, lions hunting, dogs running about and barking, youths dancing with maidens, men making war, and women cooking.

As a sight to behold, the Shield of Achilles had different effects on different people. Achilles' own retainers, the Myrmidons, cowered and would not look at the god-wrought armor, yet two of the defenders of Troy, Aeneas and Hector, who dared face the wrath of Achilles on the battlefield, were unfazed by the glamour of his shield. What was it that determined a person's response? To anyone looking at the shield, the entire world and its ways would have been revealed in an instant by the craft of Hephaestos. The terrible experience of having the weight of the world thrust upon one's shoulders would leave undaunted only the wisest, who have the intellectual courage and mental strength to bear it, and the stupidest, who lack the ability to discern the nature and extent of the burden.[2]

The story of the Shield of Achilles reaches a conclusion of sorts in Book XIII of Ovid's *Metamorphoses*. After Achilles is killed by an arrow that strikes him in the heel, an assembly of Greek chiefs tries to decide who should inherit the divine armor that he no longer needs. Only two contenders emerge: Ajax Telamon, the strongest and fiercest among the Greeks now that the great Achilles is gone, and crafty Ulysses, the protégé of Pallas Athena, the war-like goddess of wisdom. Knowing better than to try to outdo Homer, the father of all war correspondents, by describing an actual fight, Ovid imagines Ajax and Ulysses sparring over the spoils with words, not swords. The rhetorical duel that ensues is a testament to the mastery of brain over brawn, expressed in no equivocal terms. In the speech that wins him the armor, Ulysses notes, first, that his exploits and merits in the war that ended with the ruin of Troy exceed those of Ajax, and second, that the Shield of Achilles would be wasted on Ajax:

> *For that dull soul to stare with stupid eyes,*
> *On the learn'd unintelligible prize!*
> *What are to him the sculptures of the shield,*
> *Heav'n's planets, Earth, and Ocean's watry field?*
> *The Pleiads, Hyads; less, and greater Bear,*
> *Undipp'd in seas; Orion's angry star;*
> *Two diff'ring cities, grav'd on either hand;*
> *Would he wear arms he cannot understand?*[3]

Those of us who root for Ulysses in this dispute (perhaps because our own livelihood derives from brains rather than brawn) will be happy to learn that Ovid's conceit—the assumed relationship between an all-encompassing representation of the world and wisdom—has solid support from cognitive science.

We now know that the burdensome knowledge of what the world is made of and how it works *is* wisdom, in a computationally concrete sense. Insights that are continually distilled from one's accumulating history of experiencing the world, one episode at a time, keep adding to the burden of knowledge, yet knowledge that leads to understanding definitely has its payoffs. The ability to sustain the process of *learning the world* is both a precondition for practical wisdom and its happy consequence: insight breeds insight, and wisdom reveals knowledge to be rewarding, while making its burden feel lighter and the bearer happier. The cognitive faculty that makes all this possible is memory.

Remembrance of Things Past and Future

Memory imparts to a mind a kind of sophistication that can never be attained merely by being very good at real-time "war room" data processing. The mind's war room is busy at all times with the pressing need of deciding what to do next. The decisions that steer behavior take shape there through constant give-and-take among the internal representations of perceptual qualities, motivational drives, and action affordances. These intelligence items, which all originate in the more or less recent past, interact with each other and come to a head during what William James called "the *specious* present"—"no knife-edge, but a saddle-back, with a certain breadth of its own on which we sit perched, and from which we look in two directions into time."[4]

This "look in two directions" is sustained through the action of memory. Inasmuch as the representations that it deals in are momentary and fleeting, the grand map in the mind's war room is like a mirror that reflects all things as soon as, but only for as long as, they face it. Adding memory to the "mirror" stretches the mind's representational capacity from the immediate present to the past, and thereby to the future: the mirror becomes capable of reflecting also the shape of things to come.

But getting information back from the future is utterly incompatible with the foundations of contemporary physical theory, so what does remembering the future—that is, having foresight—actually mean in practice? It appears that an explanation of this bit of seemingly physics-defying magic has been discovered by J. R. R. Tolkien, who worked it into a scene from *The Lord of the Rings* that involves, very appropriately, an Elvish Mirror (with a capital M). In this scene, the Lady Galadriel offers Frodo a look in the Mirror, but warns him that

> the Mirror will also show things unbidden, and those are often stranger and more profitable than things which we wish to behold. What you will see, if you leave the Mirror free to work, I cannot tell. For it shows things that were, and things that are, [and] things that yet may be.

Just as the Mirror of Galadriel shows those who dare look into it things that *may* be, not things that *will* be, so does foresight. In full compliance with the laws of physics, foresight is never about the actual future, only about *possible* ones—or rather, the *likely* ones, given memories of the past. The magic of the Mirror is thus revealed to be merely advanced information-processing technology, whose glamour is wholly due to its computational sophistication.[5]

In a world that is predictable often enough and closely enough, the technology of foresight works by scouring memories of the past for patterns and trends that may apply to the future, turning retrospection, as it were, into prospection. Such learning from past experience happens over a range of time scales. The shortest of these may span just a fraction of a second, as when you mentally project the path of an incoming tennis ball into the future and move your racquet into a position to intercept it. The longest time scale on which learning happens is evolutionary. By driving some members of our species out of the genetic pool, selection pressure made it fractionally more likely that those who remained (a group that includes all of your ancestors and all of mine) were somewhat better at predicting whatever it was that made the difference.

Why Everything Important That You Know You Must Have Learned for Yourself

Evolution, then, is a kind of learning—the kind whose fruits are fully entitled to be called hard-won (not by you, though, if you are any good, as well as lucky). Having realized that evolution promotes the collective amassing of knowledge by various species, we may find it hard to resist the lure of drawing a crisp distinction between "innate" and "acquired" knowledge, one derived exclusively from the history of the species' interaction with its environment and the other relying on the individual's experience. Is a typical human's knowledge of the world mostly innate or mostly acquired? It may seem that after 3.5 billion years of evolution, everything about this planet that is worth learning would have already been learned and assimilated into the genome of an adequately brainy species.[6]

It turns out, however, that the very concept of innateness—let alone the idea that much of what we know is innate—is deeply and irreparably flawed. (Good thing, too, given how terminally ennui-inducing a world would be where virtually everything important that you know was already known to your parents.) One problem with it is that the folk notion of purely innate knowledge is quite impossible to pin down in rigorous scientific terms, mainly because gene expression (which happens throughout the lifetime of an organism) and cognitive development (both embryonic and postnatal) are utterly dependent on interactions between the genes, the rest of the organism, and the environment.[7]

This interaction occurs as the individual's complement of genes works together with the precision protein-building cellular machinery, which grows a body to specification and keeps it alive. Because the brain, along with the mind that it gives rise to, is part of the body, genes necessarily also help shape cognition. They do so by constraining the structure and interconnection patterns of brain areas and also, on a very local level, by constraining how synapses through which neurons connect to each other change their signal transmission properties in response to the statistics of neural activity (which ultimately relates to experience).

The "language" of connection patterns is, however, too coarse to code information that would be directly useful in generating context-sensitive future behavior from the individual's concrete past experience. For that, it is not enough to have the right neurons connected to each other: the synaptic weights at each connection must be set to very particular values. And yet, on the level of synapses, the direction of information flow in learning is the opposite of what genetic determination would predict. For example, the so-called immediate early genes kick into action in response to certain neural activity patterns (and help lay down memory traces of those patterns), rather than the other way around. All this makes the genome a very poor conduit for specific behavior-related knowledge.[8]

Even if the neurocomputational limitations on coding specific experiences did not conspire to prevent detailed knowledge from being smuggled across generations through the genome, it would still not be a good idea for a species to bet everything on knowledge that is prepackaged into a newborn's genetic endowment. Encyclopedic knowledge about the old days becomes more of a burden than an asset in a world that is changing rapidly enough. Yes, some very important properties of the universe, from the parabolicity of the trajectories of falling objects to the succession of seasons, are timelessly predictable, but other, no less important properties are subject to drift that needs to be tracked if foresight is to remain feasible.

Ironically, it is evolution itself that consistently brings about the fastest and most profound reordering. The pace of environmental change is, most of the time, glacial, compared to the constant pressure punctuated by spurts of cutthroat competition brought about by the interaction of the various species that share an ecological niche, or of the members of the same species that compete for increasingly scarce resources as it gains ascendancy and its population grows.[9]

It is usually far easier and more effective to squeeze through the genomic bottleneck traits that facilitate the gaining of knowledge rather than ready-made knowledge as such. Imagine a neural circuit that solves a particular cognitive problem by making use of knowledge encoded in its pattern of synaptic weights, each of which is a numerical representation of the strength of the "kick" that the presynaptic neuron's activity can deliver to the postsynaptic one through their connection. As I already noted, for this knowledge to be "innate," the weights would have to be dictated by the genome, raising a host of technical challenges such as installing all the right numbers in the right places.

In comparison, a learning-based solution to the same problem would require that the genome specify merely the manner in which the weight of each synapse changes in response to the activities of the neurons it connects. This extremely powerful synaptic modification rule, which relies exclusively on information that's available locally at the synapse, is actually much easier to specify in terms of an environmentally driven, genetically controlled sequence of biochemical events, making the evolution of learning from experience possible.[10]

It seems, then, that the most valuable lesson that evolution offers, to all who would listen, is that the world is inconstant but learnable and that a good living can be made by those who can learn faster than it changes. (It may also seem that not everyone need listen: if you're covered in impregnable scaly armor or are too poisonous even to look at, you may feel exempt from having to be also smart, while in truth you're just one mass extinction event away from oblivion.) Accordingly, the most valuable present that an animal may receive upon being born, or hatched, or booted up from cold storage, is the ability to learn from its own experience and from that of its peers. Like the representation of the world on the Shield of Achilles, innate knowledge may be beautiful, yet it is wholly of the past: it depicts or replays the same old scenes. The quick-witted Ulysses, who argued so eloquently that his wisdom made him the most deserving of the gift of the shield, was also the one who needed it the least. Having won it in the debate with Ajax, Ulysses did not keep the Shield of Achilles, but gave it to Neoptolemus, son of Achilles, before setting sail for Ithaca.

Mirroring the World, Mustache and All, One Step at a Time

The most literal manner in which the brain may attempt to antici-pate the future is by learning representations whose unfolding over time reflects the dynamics of the events that they stand for. Such mirroring of one dynamical system by another is not a trivial mat-ter. Unless the two systems are in every detail identical and are iden-tically connected to the rest of the world, their trajectories will sooner or later diverge, even if initially they unfold in lockstep. This representational falling out receives an exemplary treatment in the mirror scene in the Marx Brothers' 1933 movie *Duck Soup*.

As any card-carrying Marxist will tell you, in this celebrated scene Chico sets out to impersonate Groucho, with the aid of a painted mustache, black-rimmed round glasses, a fake nose, and an unlit cigar. He then encounters an identically dressed real Groucho, who decides that he is seeing himself in a mirror. (Both Groucho and the alleged reflection are wearing long white night-gowns, nightcaps, and socks.) Chico plays along by trying to mir-ror Groucho's every posture and move. The two make faces, wiggle their behinds, go on all fours, and do a couple of silly walks that may have inspired the much later opus by Monty Python. The spell is broken when Harpo, also made-up and dressed as Groucho, joins them, with predictable consequences.

The two Grouchos' aping of one another is funny because it is sustained for several minutes, because it is occasionally im-perfect, but most of all because while it lasts it appears im-probably well coordinated. They begin each silly walk while separated by a wall; they then amble, skip, and sashay in near-perfect unison across an open space where they are in plain view of each other; then another wall comes between them and the increasingly suspicious real Groucho starts plotting his next move, designed to expose his "reflection" as a fake. The fake Groucho's success in mirroring the real one's moves makes us laugh because intuitively we know just how unlikely it is that two people could execute a dance in lockstep without orchestrated timing and a thorough rehearsal.[11]

Rehearsal on the part of the system that aims at representing the world helps it shape its dynamics by giving *experience* an opportunity to make its imprint. How this happens is best understood using the no. 4 conceptual tool of representation space, from Chapter 3. Under conditions that favor learning, experiencing a series of stimuli—activating in quick succession a series of points in a representation space—causes them to become associated with one another, in the order of their activation. Repeatedly traversing such a learned trajectory consolidates it into a memory trace that is both a record of past experience and a basis for prediction. If a later event has just caused the first and then the second element in a sequence of representations to be activated, chances are that whatever it is that causes the third element to become excited will come along soon; the system now has some idea what to expect next.

Computationally, the knowledge of ordered sequential dependencies among representations has the form of conditional probabilities. This means that its acquisition and use obey the Bayes Theorem—the Promethean gift of probability theory to cognition. Once learned (through accumulation of experience, subjected to statistical inference), a pattern of sequential dependencies can be used to predict where a sequence is likely to go, given where it comes from in the space of possibilities. Like a quad on a college campus or a pedestrian square in a city after a heavy snowfall, this initially pristine representation space becomes covered with a skein of forking paths, some deeper and wider than others, which grow in response to experience.

Our encounters with the world come in fits and starts, one event at a time, with not much happening in between. Not all possible situations that could in principle be represented given the brain's resources get to be experienced, and those that do are not experienced all at once. This is what imparts to a possibility space the characteristic structure of a crisscrossing network of paths that run through an otherwise untrodden territory. The paths are punctuated with occasional stops that correspond to distinctive, hence memorable, events. At each of these stops, there is a cache of information. Because different tasks require different kinds of possibility spaces, the contents of those representational caches, as well as the pattern of paths, vary from one task to another.

The least abstract possibility spaces are those that represent aspects of actual physical space and time. To learn how to aim, or aim at, a flying object, I need to represent its possible locations and velocities. If the object is ballistic and therefore travels along a parabola, like a catapulted cow would, learning and subsequently predicting its trajectory is a simple matter of estimating a few parameters. If the object is self-propelled and has a mind of its own, like a hummingbird that flits here and there and then brakes in midair in its approach to a hibiscus flower, I need to learn and represent something of its mind.

If the object to be tracked is self-propelled as well as articulated (not the same as articulate, an attribute I'll get to in due time), and if it is acting willfully, as Groucho Marx habitually would, the problem of anticipating its moves—that is, estimating the relative probabilities of various likely future moves, given the past ones—is considerably more complex. The observer working on populating the possibility space with data must in this case learn the likely changes in the object's location (relatively easy), bodily configuration (tricky, especially if a silly walk is in the works), and mind state (very tricky, but not out of the question if you are at least as clever and observant of your object of attention as Chico Marx appeared to be of his brother).

All these learned patterns of change are represented as paths through possibility space, which I shall call PaThS (an almost-acronym that is easier to pronounce than the cryptic "PTPS" and more transparent than the pedantic "PaThPoSp"). The idea of a path through a representation space has been with us since the previous chapter, where I invoked it to explain how brains counter the effects of the changes in the visual appearance of objects, such as faces, that are brought about by changes in vantage point. As you will recall, the brain makes sense of a vaguely familiar face seen for the first time from a new angle by first populating a face representation space with tracks that record how familiar faces change appearance with vantage.

By now, it should be clear to us that the dimensions of the possibility space can stand for anything at all. Suppose that Chico undertakes to represent Groucho's state of mind by focusing on just two of its qualities: affect and arousal. For this purpose, Chico needs to allocate two dimensions of his possibility representation space—the mood plane, as it were. Within it, various conceivable mood shifts of Groucho's would be represented by PaThS of appropriate shape and placement. For example, a transition from being quietly happy (low arousal, positive affect) to being violently miserable (high arousal, negative affect) would correspond to a diagonal move from one corner of the mood plane to the opposite one.

Although most of the things and events that populate the possibility space are quite abstract, the physical environment not only gets represented within it but serves as a scaffolding that supports and structures the rest of the possibility space. The representation of the environment is privileged in this manner for a simple reason: it anchors in place episodes and actions and imposes order on sequences of events, which unfold as the learner experiences his or her or its corner of the world. As the learner gets older, avoiding being buried under a pile of indiscriminate memories becomes more and more of a challenge. If time is nature's way of keeping everything from happening at once, then space—or rather, the memory of space—is nature's corkboard, where everything that happens has its place.

Where Was I?

The evolutionary roots of the spatially indexed *episodic memory* system lie in a very common existential need. How can an animal that must leave its burrow to make a living improve the odds of getting back home at the end of the working day, preferably with provender for the family? By keeping track of what happened where on its last few forays into the wild, with an eye toward drawing generalizations about what usually, or at least sometimes, happens where (which would enable the itinerant animal to anticipate what will happen where the next time around).

Put yourself in the place of a desert bighorn sheep. You live near Borrego Springs, California. It is early August, and the heat at noon can be deadly. The neighborhood is patrolled by a puma, which, unlike yourself, is not a herbivore. Your only navigation device is your brain. If you have no idea where you are or how to get back to that water hole you drank from the other day, taking in local attractions would be the last thing on your mind, even if you love the desert as much as I do. And yet, it is the same set of cognitive computational skills that makes you good at sightseeing (and remembering the sights) and at getting back from point B to point A.

Reliable and effective way-finding, orienteering, and hoarding of personal episodic experience all depend on a representation of the layout of the environment. How does the brain do it? An effective representation supports navigation not by indicating where its owner is on some kind of mental map—if it did, bighorns would be in need of map-reading instruction and training as badly as most army cadets are. As any number of street-corner tourist maps in big cities will tell you, the answer to the question "Where am I?" is "You are here," which is true, but not very helpful. One would hope that the brain can do better than that.

Indeed, instead of being like tourist maps, which require interpretation, representations of space in the brain directly assist way-finding behavior by explicitly encoding various useful cues, such as up-to-date direction information to certain key locations. While I am confident that you know precisely where you are at this moment (*there*), I believe also that you know more than that. In particular, I bet that you can point in the direction of the closest source of potable water or food, and not just because I asked you earlier to imagine life as a bighorn. Speaking for myself, a decisive demonstration of my knowledge of where food is relative to where I am now would be for me to get up and head *directly* to the fridge, which is not visible from where I am writing these lines. (This task is not entirely out of the question: the paper-and-plaster walls of my house are no match for this grizzled specimen of *H. sapiens*.)

When the ability to take shortcuts through new territory between previously visited locations was first discovered in the rat, in the 1940s, scientists interpreted it as evidence for the existence in the rat brain of cognitive maps. Recording from the hippocampus, an area of the brain that had been implicated in navigation, they eventually discovered neurons that fired at a high rate only when the rat visited particular locations. These "place cells" serve as the foundation for representing space—and, it turns out, much more.

Location cues derived through dead reckoning from wandering about are enough to form the hippocampal representation of space: much of rats' way-finding is done in the dark of the night, and even blind mole rats (such as Molly from Chapter 2) start taking shortcuts across the open space in the middle of an enclosure after exploring its boundaries. At the same time, any additional cues that happen to be available get tacked right onto the basic spatial scaffolding. Thus, the so-called place cells respond selectively not just to particular places but to the combination of sights, sounds, smells, and textures encountered in those locations—in other words, to episodes that the animal experienced there.[12]

The varieties of episodic memory maintained by a species depend on its habitat. For a desert bighorn, this has always meant just a patch of territory. We humans started that way too, but have by now assimilated into this category various spaces that are not entirely, or even not at all, spatial. Instead, these spaces are abstract, like the possibility space. In such a space, one can still meaningfully discuss proximity (which translates into similarity between abstract objects, such as two action plans). Even for us, however, the roots of abstract representations are firmly planted in the basic functional need to explore and make sense of the wild environment into which we are born, armed only with the knowledge of how to learn and a notion that every thing has its place.

For human babies, forays into the wild begin in the playroom or in the backyard as they crawl about and explore the little world centered on the place where they are released by a caregiver. Among the kinds of regularities that babies need to discern in the maelstrom of sensory information that spins around them are names for things—certain sound or gesture combinations that go with certain objects or actions. Sometimes objects do receive explicit labels; only inconsiderate or stoned parents would ever exclaim, "Look at the bunny!" while pointing at something other than an actual bunny (live or stuffed). More often than not, however, the baby has only circumstantial evidence to go by, as when an object gets both mentioned and shown, but not at the same time.

Experimental studies show that in such cases babies learn more reliably if the naming of an object and its appearance are made to share the same physical location—for instance, by offering the baby a verbal label for a novel object while snapping fingers in the place where it had appeared earlier. In using space in this manner, human babies very likely rely on the same brain circuits and mechanisms that support location-bound episodic memory in other mammals (all of which have a hippocampus). It is because of this sharing of dedicated computational resources between episodic memory and language-related tasks that drivers are more likely to get lost if they are made to navigate a not entirely familiar city while maintaining a conversation.[13]

A Moveable Feast

Not only mammals qualify for an episodic memory system: chickadees, nutcrackers, jays, and titmice have it too. These and many other species of birds depend for their survival on caching food items, such as seeds or dead worms, that they later retrieve. The number of caches typically runs in the thousands, underscoring the large memory capacity that is needed to support this behavior. Carefully controlled studies have revealed that this memory is episodic: rather than relying exclusively on common characteristics of a cache location, birds memorize the actual locations they visit. They also remember the type of the food item that they store in each location: scrub jays, for instance, return to caches of perishable wax worms before revisiting places where they have stored pine nuts, which do not spoil.[14]

Because of the usual selection pressure, the ability to memorize past episodes is indulged by evolution only insofar as it carries future dividends. The scrub jay's obvious future payoff from remembering where it cached a wax worm is a tasty snack at a later time, but there is also a less obvious and much more interesting side to episodic memory, avian or human: it can support time-travel, of a kind that is perfectly compatible with the laws of physics and fully paid for by the evolutionary benefits it confers.

It is easy to see how a mind that is equipped with episodic memory has what it takes to travel mentally into the past: it can do so by recalling the circumstances of previously experienced episodes and re-creating them in the represented present, within the workspace of the mind's war room. With only a slight modification, the very same set of cognitive tools can also support mental travel into the future: one needs merely to modify certain aspects of the represented situation to turn retrospection into prospection.[15]

The study that demonstrated this ability in scrub jays took advantage of their gourmand predisposition. Jays prefer not to eat the same food day in and day out if they can help it. To motivate them to think about a future meal, experimenters taught the birds, over the course of a few days, that in one of two locations they would find a breakfast of peanuts and in the other—

kibble. When given a choice of food to store on the eve of the test day, the jays cached kibble in the peanuts-for-breakfast location and vice versa, thus demonstrating not only a love of dietary variety but also an ability to indulge it by anticipating and acting upon a future need rather than an immediate urge.[16]

In this task, the key aspect of the represented breakfast location that the bird's brain needs to modify to switch between the present and the future is its food label. In the "present" setting, the food cue is simply what the perceptual system tells the rest of the brain it is. In contrast, in the "future" setting it is what the bird expects, or would like it to be, given its earlier experience with the location. It is easy to envisage how switching between these two settings can be accomplished by neuronal circuits in which "place cells" act in tandem with neurons that represent various food concepts.

Systems based on episodic memory explore the future by reimagining the past, just as in way-finding they plan future moves by consulting past peripatetics. In the rat brain, episodic place-specific cells, whose response properties are shaped by the animal's situated experience, can also fire prospectively as the rat ponders the choice that would bring it to the location in question. The circuits that encode episodic information and thereby allow rats to weigh future options are consolidated during sleep as the animals dream about the places they visited during the day and rehearse their actions.

This is very likely the same neural system that helps humans navigate, not just through "space" space, but also through what I called earlier the possibility space, in which sequences of complex actions are learned, planned, and executed (as demonstrated with such flair by Chico Marx's anticipation of Groucho's every move in front of the nonexistent mirror). But does exploration of possibility spaces really count as mental time-travel? One could argue that planning for the future may in principle proceed "functionally," by taking estimated future needs into account, without ever leaving the experienced—or, to use a philosophical term, *phenomenal*—present.

What distinguishes a truly experiential, phenomenally prospective state from one that pertains to the future only functionally is spatial grounding. A first-person perspective—the feeling of being right here, right now—is a key component of everyday phenomenal experience (about which I'll have more to say in Chapter 6). This feeling is easy to manipulate, as the following two little experiments will demonstrate. (Don't attempt them if instead of reading this book you are listening to it while driving.)

First, think of your fridge; now open it mentally and tell me whether there is any milk left. Is there? If despite my warning you did try this experiment while driving, and if you are still in one piece, slow down and listen to any noises originating from underneath your car: while mentally sticking your head inside your fridge, you may have run over something that you should have braked for.

Now, imagine that a freak storm has buried your neighborhood under deep snow and think of opening your fridge in two days' time. Is there any milk left now? Your answer is likely to differ from the previous one, because things have changed: the kitchen looks strangely bright because of all the snow outside, and the milk carton feels much lighter. This stands to reason: you are in the same imagined place now as before, but at a different imagined time.

These exercises (easy, fun, and safe, if you followed the instructions) illustrate the phenomenal realism of mentally shifting the first-person perspective from here and now to there and then.

This makes mental space-time travel real in the most important relevant sense: for the experiencer, who necessarily always has the last word in matters pertaining to his or her or its own experience, it *feels* real enough.[17]

Because your space-time machine is inside you, rather than the other way around, it works only for you. Only you can use it, and then only to travel to your own (imagined) past or future, or, if you fix the time dial and spin the space dial, your own (imagined) geography. Because there must be a "there" and a "then" for you to go to before you can leave the here and now, the more you have been around (in time and space), the richer the virtual universe that you can explore. Most importantly, because this exploration in turn enriches your cognitive arsenal, making you better fit to face the *real* real world, it is viewed favorably by evolution.

The evolutionary angle helps us understand not only why mental space-time travel comes to us so easily, but also why we indulge in it at every opportunity—to the extent that special training in awareness meditation is needed to overcome the virtual wanderlust.[18] Episodic memory mechanisms make it possible for the hedonic value of a happy experience to transcend the boundaries of real time and space. This is why reminiscing about past happy experiences or anticipating future ones may feel good—under the right circumstances.

As Dante notes in *The Divine Comedy*, "There is no greater sorrow / than to be mindful of the happy time / In misery."[19] By the same account, however, happiness is brought into sharper focus when experienced against a dark background. In Book XV of *The Odyssey*, the old man of Ithaca who tends pigs for the household of its long-absent king befriends a beggar, newly arrived on the island after much ill luck and hardship. Having heard his complaint, the wise swineherd encourages the beggar to enjoy the offered food, wine, and shelter, because, he says, "In later days a man / can find a charm in old adversity, / exile and pain."[20]

The swineherd does not recognize in the beggar Ulysses, who has secretly arrived back in Ithaca after twenty years of war and peregrinations. The genre of the novel having not yet been invented, Homer does not tell us what it was that Ulysses felt during his long exile, but it seems safe to assume that what kept him going was the promise of happiness—the anticipation of the much-hoped-for return home.

Following the sensibilities of the age, the story does not suffer Ulysses to return home empty-handed. The fabled Phaiakians, who ferry him overnight from their island to Ithaca and set him, still sleeping, on the shore, leave him many gifts: cloaks and tunics, cauldrons and tripods, bronze swords and golden winecups, and (somewhat improbably) bars of gold. Everything that we hear about Ulysses tells us, however, that apparel, accessories, and even gold have no dominion over him. Ajax, having lost the debate over the Shield of Achilles, goes mad with thwarted greed and kills himself; Ulysses, having won it, happily gives it up.

Recent empirical studies of the comparative hedonic value of various kinds of experience suggest that Ulysses was wise in treating material possessions lightly. The enjoyment of material purchases tends to fade with time: the kick you get out of becoming the owner of a new cooking cauldron is soon gone, even if satisfaction from its use persists. In comparison, experiential purchases leave a trace of happy episodic memories that can be relived: go sailing around the Mediterranean and you'll appreciate the thrills of that experience for the rest of your life, even if it takes you a decade to get back home.[21]

As Ulysses arrives at long last in Ithaca, he is fabulously rich, and not because of the Phaiakian fortune. Among his real treasures are his memories of the victory at Troy and of the long and circuitous journey home. Listen to Constantine Peter Cavafy, the melancholy Greek poet of exile:

> Keep Ithaka always in your mind.
> Arriving there is what you are destined for.
> But do not hurry the journey at all.
> Better if it lasts for years,
> so you are old by the time you reach the island,
> wealthy with all you have gained on the way,
> not expecting Ithaka to make you rich.[22]

It is all good, down to the last line, where, it seems to me, Cavafy, who lived and died alone in distant Alexandria, speaks out of bitterness. Ithaca does make Ulysses rich, with the priceless treasures of home: the sound of the surf in its coves, the texture of its stones under his sandals; the smell of its wild thyme; the sight of its olive trees and vines; the taste of their fruit; and also: the approval and admiration of his father, Laërtes; the valor and presence of mind of his son, Telemakhos; and the companionship and love of his spouse, Penelope.

SYNOPSIS

Not just cognition but life itself, which depends on biochemical information processing, would be impossible were it not for the world's predictability. Yet, the patterns formed by many types of events that matter to a mind that is trying to fend for itself, such as competition among individuals and species, shift over time. Because of that, and because of how difficult it is to squeeze specific helpful hints about how the world works through the informational bottleneck of the genome, evolution's best bet is on minds that can learn through experience and thereby both attune themselves to environmental regularities and deal with environmental change.

What kind of computation does it take? To discern regularities that can be relied upon in planning future behavior, the brain must keep track of past events and the sequences they form and distill from them trends that can be projected into the future. This insight reveals the true computational task of a faculty of the mind that is familiar to all of us as memory. Far from being a mere repository for odd pieces of information, your memory is charged with relating the episodes of your life to each other, seeking recurring patterns—crisscrossing paths that run through the space of possible perceptions, motivations, and actions.

Because of its likely evolutionary roots in way-finding, episodic memory relies heavily on taking note of locations in which events happen and of their spatial relationship, turning a representation of the layout of the physical environment into a foundation for the abstract space of patterns and possibilities that it constructs over the mind's lifetime. As they fall into place the paths through the possibility space can support mental travel in space and time, which are both simulated to the best of the mind's knowledge and ability. Episodic memory is thus the mind's personal space-time machine—a perfect vehicle for scouting for and harvesting happiness.

5 | You Can Talk to Me

Preeminence above a vole. Replicants abroad.
The digital revolution. Reduce, reuse, recycle.
A garden of forking paths. Dependencies all
the way down. It takes a village.

You can talk to me.
You can talk to me.
You can talk to me.
If you're lonely, you can talk to me.
   —THE BEATLES, "Hey Bulldog" (1969)

The Garden of Forking Paths is an incomplete, but not
false, image of the universe as Tsúi Pên conceived it.
   —JORGE LUIS BORGES,
      "The Garden of Forking Paths," *Ficciones* (1941/1962b)

Preeminence Above a Vole

The Odyssey, as one of its best translators, Robert Fitzgerald, noted
in a postscript, is "about a man who cared for his wife and wanted
to rejoin her." For three thousand years now, this story captivates
listeners and readers not just because it brings the hero through
spectacular hardships to a happy end, but also, perhaps, because it

presents a puzzle. On several occasions during his ten-year journey home, Ulysses could have renounced adventure and avoided further adversity by settling down with one of the many willing women he met, some of them, like Nausikaa, mere kings' daughters, others, like Kirke and Kalypso, immortal. Why did he persist?

Young Nausikaa, admiring Ulysses as he passes on his way to the feast in his honor in the hall of her father the king, simply bids him remember her when he is safe at home again. The goddess Kalypso on her island, though, demands an explanation when it is her turn:

> Son of Laërtes, versatile Odysseus,
> after these years with me, you still desire
> your old home? Even so, I wish you well.
> If you could see it all, before you go—
> all the adversity you face at sea—
> you would stay here, and guard this house, and be
> immortal—though you wanted her forever,
> that bride for whom you pine each day.
> Can I be less desirable than she is?
> Less interesting? Less beautiful? Can mortals
> Compare with goddesses in grace and form?

> To this the strategist Odysseus answered:

> My lady goddess, here is no cause for anger.
> My quiet Penelope—how well I know—
> would seem a shade before your majesty,
> death and old age being unknown to you,
> while she must die. Yet, it is true, each day
> I long for home, long for the sight of home.

As this conversation winds down, Kalypso, who has been ordered by Zeus to let go of Ulysses and help him sail home, promises to get to it first thing in the morning. That she does, but not before they do one more time what they have been doing every night for years, ever since Ulysses was washed ashore on Kalypso's island: they go to bed.

Although the idea of an undying beauty being let down and left flat by a mortal is bound to resonate with any human audience, our enjoyment of the story would be incomplete if the mortal's motives were entirely opaque to us. That they are not: those of us especially who know what it means to have a home and a soul-mate have pretty good intuitions as to why Ulysses spurned the offer of immortality in favor of returning to Ithaca and his Penelope.

The mind of Ulysses cannot be understood solely in terms of abstract principles, no matter how large these loom in Kalypso's calculus of desire. A side effect of the Greeks' creating their gods to personify basic drives—the just Zeus, the angry Poseidon, the war-like Ares, the smutty Aphrodite—is that the entire Pantheon sitting in caucus may find it hard to fathom a mortal soul, especially if Athena is on vacation. By making Ulysses an offer he can refuse, the nymph Kalypso thus acts very much in character.

Kalypso's bid would have better chances for success in a Beatrix Potter universe, with all the humans replaced by rabbits, frogs, or perhaps talking fish. In all vertebrates, social behavior traits such as gregariousness and monogamy are shaped by balancing the same few endocrine factors. For instance, the relative abundance of the hormones oxytocin and vasopressin in certain key areas in the brain determines whether a species is monogamous or not. This is why prairie voles that lose a mate never take on another one, whereas montane and meadow voles sleep around like it's the 1960s.[1] One suspects that an account of a prairie vole's odyssey on his way back home from a war instigated by the misbehavior of a bunch of meadow voles would be an epic bore.

In humans, the standard-issue vertebrate social behavior network is topped with a huge information-processing apparatus, which makes life interesting. Equipped with a brain that has computational capacity to spare, versatile Ulysses is neither slavishly monogamous nor willfully, or even just willingly, promiscuous. Behind his happiness with Penelope (who is often called in *The Odyssey* περίφρων, "very thoughtful"), there is another power, which deals in information. This power is an active informational *being*, which lives in symbiosis with humans but not with voles or any other species on this planet. To such beings we rent out real estate in our brains in exchange for assistance in love, war, and general procurement of fun. The product of the symbiosis between these beings and our brains is language.

Replicants Abroad

Information, as everybody knows, wants to be free—which may sound like a pipe dream, given the fondness of human institutions and societies for corralling and policing it. One of the few glimmers of hope for the freedom of information is for it to assume the initiative. When chunks of knowledge turn out to be capable both of propagating themselves and of adapting to the quirks of their hosts, information comes alive.

The meaning of "life" being simply self-replication with occasional heritable changes, there is no difference to speak of between a chunk of information that causes its current host to make it available for pickup by a new host, and a virus, which hijacks the transcription machinery of the infected cell to make copies of itself and release them into the wild, where they can infect other cells. The word for such a live chunk of information, coined in 1976 by Richard Dawkins, is *meme* (pronounced so as to rhyme with "gene").[2]

A perfectly good, if not entirely elementary, example of a meme is *The Odyssey*—a complex sequential pattern of information with a long lineage. Its many ancestors, all slightly different from each other, originated and survived in the brains of roaming rhapsodes. With the invention of the new cognitive technology of writing, which allowed memory to be outsourced, one of the versions committed to paper became canonical.

Its continued appeal to members of the host species ensures that the *Odyssey* meme is allowed to replicate, with the number of copies having long ago surpassed the number of people who ever had a chance to hear it sung. Although its very popularity caused it to become immutable, it keeps spinning off other, no less complex, memes that are anything but limited in their content and style, ranging from James Joyce's *Ulysses* to Zachary Mason's *The Lost Books of the Odyssey*. There have even been some attempts to harness Homer's heroes to the chariot of popular science.

Standing back from these examples, one realizes that the idea of borrowing from the classics to bolster one's own writing is now itself a meme (a love child of plagiarism and homage, as it were). This little discovery in turn suggests that the concept of "meme" is a meme too, one with which I may have just infected you—unless, that is, you have already been infected, or else are highly skeptical and hence immune, skepticism being the mind's first line of defense against virulent concepts.

Does this mean that there is a meme behind every exchange of information out there? To distinguish between empty meme-talk and a genuinely explanatory conceptual move, we should ask who benefits from the transaction in which meme involvement is alleged. If the suspected meme clearly profits from what is going on, by becoming more successful in its bid for self-propagation, the memetic explanation stands. This is true even if the host, other memes, and other hosts benefit too: unlike the giving of tangible goods, passing on information does not imply having to part with it. A meme's involvement is, however, particularly strongly suspected if everybody else finds themselves inexplicably worse off when the dust settles. An extreme example of such a selfish meme is an information pattern that codes for a polarizing, proselytizing, martyrdom-endorsing religion, which can cause you not merely to try to infect other people with it but to die trying, just as it settles comfortably into the brains that became available to it through your efforts.[3]

The least that a religion bug could do for budding martyrs would be to make them happy while they wait for their chance, but even such a halfhearted handout to the host is by no means commonplace. The memes that you put up in your memory space care about your happiness no more than you care about public transportation in your hometown. Sure, it would be nice if it were easier to catch a cab when it rains, or if the trains on your line ran more frequently, but it would be odd for a commuter to become emotionally attached to cab number 1729 or to a particular train on the F line. (The public transport analogy is imprecise because, as noted earlier, you do not leave a copy of yourself in every cab you ride in, but it is otherwise perfectly applicable.)

Because for them you are just a commodity, there is considerable variance in the effects that memes end up having on your subjective well-being. A meme may make you happy, as when you learn that all you need is love (provided that love is indeed there to be had). Or it may leave you indifferent, as when you get to know the multiplication table (a useful skill, but not a particularly exciting one, either way). It can also make you pretty wretched, as when you succumb to the belief that your life is being secretly manipulated by the Illuminati of Bavaria, or when you catch strict Calvinism.[4]

Nonrandom behaviors, such as deciding that all you need is love and then acting on it, are shaped by memory representations (through an interaction with the environment). Because of that, the spread of a behavior through a population of hosts is a sure sign that the meme that codes for it has replicated itself many times over. As they propagate, memes may undergo mutation because of imperfect pickup, retention, and reproduction of behaviors: an off-duty milkman overhearing the Sermon on the Mount from the next hill over may leave with a firm conviction that it is the cheesemakers who are blessed.

How widespread a given meme eventually becomes in a population depends, in each case, on the processing of information by the hosts and on the dynamics of contagion. As in regular epidemiology, this includes a multitude of factors, from the allure of the meme and the susceptibility of the host to the quality of the environment that the host offers and its ability to reproduce faithfully the behavior in question. A catchy tune or a lifestyle fad may sweep through millions of brains in a few months, or it may fade into the long tail of the infosphere's popularity chart and quietly proceed to dwindle into nothingness.

A meme would seem to stand a better chance if its replication happens to be tied to the actual primary use of the corresponding behavior by the host. A handy (or rather, beaky) example is found in the evolution of tool use in New Caledonian crow populations, which is cultural, not genetic. It begins with a minor eureka moment: a crow discovers how to shape a screw-pine leaf into a tool, then uses it to extract insects from rain-forest vegetation. Another crow observes and learns, either by imitation or by informed trial-and-error. After some time a population of interrelated memes—variations on the common theme of how to make a tool out of a leaf—has taken over the crow population.[5]

Memes such as memory blueprints for birdsong, whose behavioral manifestation *is* signaling as such, are ideally positioned for insinuating themselves into the memory of additional hosts. Male songbirds sing to impress the females (which in most species remain demurely silent). Adolescent males learn the repertoire of their species (usually quite limited, but open-ended in a few species, such as the sedge warbler or the superb lyre bird) by listening to adult tutors and doing their best to sound like them, while keeping an eye on social feedback from females. In this manner, the bird population coexists with a population of songs.

Just as eking out an existence in the state of nature is always subject to evolutionary pressure (birds competing over mates; song memes competing over memory space in the birds' brains), so the *co*-existence that is inherent in the meme-host relationship is subject to *co-evolution*. The push and pull of co-evolution is particularly powerful when this relationship is mutually beneficial (symbiotic rather than neutral or parasitic), as it is in birdsong. Male birds who sing better get more chances to dazzle a female, leading to the genetic evolution of brains that are better at learning, retaining, and performing songs. At the same time, songs that better fit the existing bird brains and vocal apparatus get more chances to replicate, leading to the (often much faster) cultural evolution of songs that are easier to learn, retain, and perform.

In no other case has co-evolution of memes and their hosts led to a more world-shattering outcome than in human language.

In this one species of perpetually hungry, highly social, and highly competitive information processors, which threw in its fate with an initially small band of information packets that proved highly infectious, selection pressure set off a runaway cascade of trans-generational cultural learning. The still somewhat bewildered beneficiaries of the resulting relentless series of cognitive system upgrades—some genetic, but many more others cultural—now find themselves collectively capable of working miracles, such as killing millions of their conspecifics at the press of a button, saving millions by inventing antibiotics, wrecking their home planet, and landing fancy hardware on other planets.[6]

We come to share in the cognitive kickbacks of language because learning it, in all its fantastic complexity, is for us mere child's play. Indeed, language itself is a kind of game that all of us play—a structured activity in which we engage for fun or profit, together or alone.[7] It is, however, a peculiar game. Participation is not a matter of choice: normally developing children get sucked into the language game simply by being their regular social selves around their peers and other people, who already are in the play. Quitting the game once you're in is not an option either: you can stop speaking to other people and you can stop your ears, but you cannot stop interpretations of what you hear if you do listen (or, in sign language, interpretations of what you see if you do look) from arising in your mind.

Taking the memes' perspective, we see that language is also a game that plays people. The memes that comprise it are not, after all, passive packets of information. The thoughts and expressions that act as pieces in the language game have a mind of their own, in which they resemble the hedgehogs that served as balls in the Queen of Hearts' croquet game in *Alice's Adventures in Wonderland* and that had a predilection for moving around without waiting for Alice to strike them with her flamingo. Meme-instigated thoughts hover in the background, intervening whenever possible in actual overt behavior, constantly on the watch for opportunities to express themselves so as to be seen or heard by other potential hosts. A day on which a meme gets its human host to air it is a happy day for it and for the memories of the words and gestures it recruited along the way, all of which receive a nice representational health boost from the exercise.

The word croquet game is by orders of magnitude more complex than any of our other ritualized activities that do not rely on language. The troupe of memes that collectively turn a human into a language player grows well into tens of thousands of active memory traces of all stripes and sizes. Together they impart a ritualized—statistically regular, hence meaningfully learnable and sharable—form to the thoughts that well up and make themselves available for sharing. With a bit of experience and mental agility on the part of the team (the troupe and its host), the outcome of this process can be breathtaking, as the following dialogue amply illustrates:

ROMEO, *to* JULIET

   If I profane with my unworthiest hand
   This holy shrine, the gentle fine is this:
   My lips, two blushing pilgrims, ready stand
   To smooth that rough touch with a tender kiss.

JULIET

   Good pilgrim, you do wrong your hand too much,
   Which mannerly devotion shows in this;
   For saints have hands that pilgrims' hands do touch,
   And palm to palm is holy palmers' kiss.

ROMEO

Have not saints lips, and holy palmers too?

JULIET

Ay, pilgrim, lips that they must use in prayer.

ROMEO

O, then, dear saint, let lips do what hands do;

They pray, grant thou, lest faith turn to despair.

JULIET

Saints do not move, though grant for prayers' sake.

ROMEO

Then move not, while my prayer's effect I take.

Thus from my lips, by yours, my sin is purged.

> ROMEO *kisses* JULIET

JULIET

Then have my lips the sin that they have took.

ROMEO

Sin from thy lips? O trespass sweetly urged!

Give me my sin again.

> ROMEO *kisses* JULIET *again*

JULIET

You kiss by the book.[8]

The Digital Revolution

Like a soccer match overrun by spectators who spill over onto the field, chase away the referee and the teams, and tear up the turf, a language game without order and structure would quickly devolve into verbal chaos. There being no pattern in mayhem for a novice participant or an outside observer to discern, a game that is prone to disorder is unlikely to survive as a meme: such games become extinct just as soon as they coalesce out of the background noise of their hosts' behavior.

Whereas in a disorderly game each melee is chaotic in its own inimitable (and therefore unlearnable) way, in a well-structured game all rounds resemble each other. All instances of the game of Go are played out with pieces that differ only in their color; all soccer matches involve one ball each; all wedding ceremonies decrease the number of unmarried people by no more than two at a time. In language use too, there are certain structural traits that hold for all conversations and for the language game in general.

The most obvious such trait is the serial order in which language is generated and perceived. In spoken language, it reveals itself in the sequential structure of speech: sounds follow one another in an order that matters, often accompanied by a series of gestures and facial expressions that are likewise ordered and timed. Meddling with the order of the sounds is generally a recipe for total communication breakdown. Meaning may merely mutate, as when "kiss" pronounced backwards becomes "sick," or as "you kiss by the book" may be turned into "you book by the kiss" by a spoonerism-prone novice Juliet, gripped by stage anxiety. Meaning is far more likely, however, to vanish altogether: an overwhelming majority of conceivable sound combinations are not just meaningless but unpronounceable.

Although the ordering of words is what first comes to mind when one thinks about its sequential nature, language is serial on more than one level. Sequences of basic sounds or "phones" form words, which in turn can be strung one after another to form phrases and sentences. As phones are produced and perceived by the members of a linguistic community who share a dialect, they are channeled into a small number (no more than a few dozen) of distinct categories—the phonemes. All human languages, as they are spoken (or signed), are in this sense digital: they are constructed from discrete building blocks in the same way that the file into which I am writing these words exists in my computer's memory as a sequence of physical symbols for 0 and 1.[9]

Getting the phonemes right may be quite a tough job for a non-native speaker. Having been brought up speaking Russian, I, for one, will always be challenged by some of the phonetic distinctions that English mandates. (The need to avoid making tricky distinctions explains my otherwise puzzling tendency while speaking English to substitute whenever possible "page" for "sheet" and to refer to the German philosopher Kant, with whom I never actually drank bruderschaft, as "old Immanuel.") At the same time, I am told that in Hebrew, which I speak as fluently as any native, I sound like an American.

In contrast to inept foreigners and their queer phonemes, both the native speakers and the language memes they bandy about benefit greatly from the categorical limits on phonemic variation. For the speakers, a digital medium affords error-free communication; for the memes, it promotes faithful replication. It was by going digital that the language memes adapted to their human hosts' predisposition for perceptual categorization, while offering them a handsome kickback in the form of a reliable medium for communication. This co-evolutionary development set both sides in the language game on a path to their unprecedented collective success.[10]

Reduce, Reuse, Recycle

The digital revolution in cognition would have not gone far were it not for another key structural trait that is common to all languages: arranging and rearranging the same digital building blocks to construct a potentially unlimited variety of complex messages, by the way of constrained combinatorial composition. By themselves, digital memes are useful (and fecund) because they are easy to trade. Assigning a distinct symbol to each thought, however, would quickly exhaust the cognitive resources of even the brainiest species. Much worse, it would rule out any possibility of communicating a thought for which the originator and the intended recipient do not already share a symbol.

Both these problems can be averted by the same means: *reuse* of partial structures in different contexts, which includes constructing larger novel structures out of smaller existing ones. Under such a recycling scheme, a part's contribution to the meaning of the whole (that is, to the effect that it has on the listener) depends on its context. Because different combinations of familiar parts can mean different things, novel meanings can be expressed in a form that would not leave the listener too bewildered. The tale of Queen Mab that Mercutio spins for Romeo just before they crash the Capulets' party no doubt transcends the prior linguistic experience of that self-confessed "lusty gentleman"—

MERCUTIO
   She is the fairies' midwife, and she comes
   In shape no bigger than an agate-stone
   On the fore-finger of an alderman,
   Drawn with a team of little atomies
   Athwart men's noses as they lie asleep.
   [thirty-seven more lines of wild fantasy omitted]

And yet, Romeo finds Mercutio's ravings not so much incomprehensible as perhaps just slightly boring.

ROMEO
  Peace, peace, Mercutio, peace!
  Thou talk'st of nothing.

Whatever Shakespeare's characters have to say about their author's wordplay, his audiences delight in it. (I still remember being
electrified by the Queen Mab rant, delivered by John McEnery in
Franco Zeffirelli's *Romeo and Juliet*, which I first saw in the early
1970s.) Why is it so?

In his 1765 *Preface to Shakespeare*, Samuel Johnson remarked
that "the dialogue of this author . . . is pursued with so much ease
and simplicity, that it seems scarcely to claim the merit of fiction,
but to have been gleaned by diligent selection out of common conversation, and common occurrences." The delight that we take in
Shakespeare's words (or in those of Homer) is a close relative of
our pleasure in language in general, whose roots go to the core of
what it is to be human.[11] Our happy symbiosis with language ensures that we use it often, and with relish; the skill of a master of
dialogue or narrative serves merely to amplify the features that are
present to some extent in any spirited conversation or good story.

As in the rest of cognition, the pleasure in language is derived
from a properly maintained mix of familiarity and novelty—
something old and something new. On the one hand, familiarity
with at least some parts or aspects of the stimulus ensures that
the perceiver's brain circuits will not remain indifferent to it.
Novelty, on the other hand, promotes an effort on the part of the
brain to make sense of the stimulus. This effort, if not thwarted
by too much that is strange, allows the mechanisms that carry it
out to justify their upkeep and reap the reward of pleasure.[12]

What seems strange to a mind and what does not depends, of
course, on how many of the world's gifts it has sampled. It would

be wrong to think of worldly experience as a hoard of information that the mind may consult or ignore at will. Memories of experience—from simple sensory impressions to sophisticated speech acts—become part of the mind and thereby irrevocably transform it. We may smile while reading Jorge Luis Borges's "Happiness," which opens with these lines—

> Whoever embraces a woman is Adam. The woman is Eve.
> Everything happens for the first time.

—yet as the spell cast by the poem passes, we recognize that a person can look at the world with new eyes, either by setting aside the burden of memory or by accepting it, only at the cost of becoming someone else.

Such change need not be shunned, nor should it be embraced, without due consideration. The dynamics of remembering and forgetting and its interplay with the affective value with which memories are often suffused are too complex to be given proper treatment here, but we definitely have the conceptual tools needed to ask the right questions. For one thing, we know that insofar as memories and emotions are part of what a person is rather than something that a person owns, any consideration in this matter will be intensely idiosyncratic.

A chief argument in favor of remembering as much as possible is the opulence of nuance and association that memories bring to ongoing experience. In my mind, this great gift of memory far outweighs the advantages of selective forgetting, which is usually attempted as a remedy for remembered pain. As suggested by studies of subjective well-being, the negative affect associated with a memory fades rapidly with time. The same, however, goes for positive affect, which is why happiness cannot be saved and stored for future use.[13] With past affect all but gone, what remains behind and endures is the cognitive essence of condensed experience—a lasting treasure that, like the burning bush in the Sinai wilderness, gives light without being consumed.

In a recent desert experience of my own, I was given a chance to appreciate the potential of remembering things—in this case, some well-chosen words—for helping one see one's surroundings in a new light. I was walking up a gradually narrowing dry wash, leading into the very fittingly named Smoke Tree Canyon, in the California desert, when the smoke trees' halfhearted attempts to snag my sleeves with their thorns brought to my mind, of all things, a line from Borges's "Happiness": "The calm animals come closer so that I may tell them their names." I enjoyed this thought so much that I sat down for a few minutes in the shade of the canyon wall to savor the view that evoked it. As I was getting up I saw that I had almost sat on a beautiful, pencil-thin, black- and orange-banded snake—which I could not, unfortunately, inform of its name because I did not know it at the time.[14]

A Garden of Forking Paths

In playing with words, more than in any other cognitive domain, the process of gaining experience by assimilating an utterance or generating a new one is constrained to visit a particular sequence of categorical representations—digital units such as phonemes or words. The resort in language to sequentially ordered associative memory showcases the spirit of functional thrift that permeates cognition: never invent a new solution to a problem if an existing solution to another problem can be adapted instead.[15] In Chapter 4, I called this solution PaThS (paths through possibility space). Because in language the space of possible strings of representations (sequences of units) is discrete, it is best visualized not as a field across which one can blaze a trail any which way, but as a set of stops connected by a rail network.

The thing to visualize is, moreover, not some perfectly ordered railroad yard (say, the new Berlin Hauptbahnhof), with immaculate tracks and an ironclad schedule. Imagine instead a loopy maze of crisscrossing tracks that may have gaps, dead ends, and seemingly redundant parallel segments—the subterranean labyrinth of cart tracks running through the network of mining tunnels in Steven Spielberg's *Indiana Jones and the Temple of Doom*, prone to all manner of sabotage and beset by avalanches, floods, and occasional lava intrusions, the equivalents of various common irregularities of natural speech, such as disfluencies or utterances that get abandoned in midsentence.[16]

The stops along these tracks are words; the routes that carts take through the system spell out longer utterances. The selection of words and their order within the utterance are constrained by the labyrinth's layout, by the existing connections, by the settings of the switches, and by whatever repair work or volcanic eruptions may be in progress. The point of language being usually communication, the chosen words and the order in which they appear are also constrained by the speaker's intent, which the utterance can no more than hint at.[17]

In choosing and stringing together the words for this last paragraph, I have been very, very careful, lest too much of my intent get lost in transit between my brain and yours. I avoided using the word "meaning," which is infamous for being as difficult to define as it is easy to misconstrue. (I shall return to it a bit later, when it is safer to do so.)[18] Instead, I wrote "intent," about which people's intuitions tend to be generally on the right side of progress in cognitive science. I also made a point of noting that utterances proffer clues about intent, rather than ferrying explicit content.

What, then, is this intent that shapes linguistic behavior? The seas around the shoals of semantics, the discipline charged with answering such questions, are bestrewn with wrecks of theories that foundered under the weight of their own mathematical armor. To traverse these treacherous seas, we need a craft that sits light in the water and responds well to the breezes of common sense—not a battleship but a dinghy. Let's see if we can build one out of the materials at hand, using the conceptual tools accumulated over the course of the last several chapters and keeping in mind in particular evolution's knack for recycling. Perhaps a computational stratagem that works elsewhere in cognition is involved in implementing intent too.

A particularly handy and versatile tool in our kit is the idea of distributed representation that takes the form of coordinated activity of multiple units and circuits in the brain. It would make sense for language to be similar to visual or auditory perception, where a face or a sound can be represented by the activities evoked in a set of units tuned to other faces or sounds, as well as to motor control, where a novel action can be generated by interpolating the activities of units tuned to a few select members of the space of possible actions.

The intent behind an utterance is, then, a surge in the activities of a "cloud" of representations in the mind of the speaker. The immediate origin of some of these activities is external, in that they pertain to various perceptual aspects of the speaker's present situation; other activities arise internally, having to do with memory (both retrospective and prospective), simulated actions, and motivation. Eventually, the dynamics of communicative intent transforms this activity into motor commands that deliver into the world an externally observable act—the only medium through which the memes behind the intent can reach another mind.

As the utterance hits the listener's sensory apparatus, it evokes an interplay of representations with very similar dynamics[19] that involve perceptual representations, predictions based on episodic memories, simulations aimed to evaluate possible courses of action, motivational accounting, and the ubiquitous emotional effects. This ensemble activity is the whole point of the communication event that just unfolded—in fact, it is now safe to reveal that this is the meaning that got across. There is no need for further interpretation; the buck, as it were, stops here. Pressing the question of the meaning of "meaning" any further would be like calling for a proper definition of a dance while insisting that it is not the same as the sum total of the coordinated motions of the dancers.

All this computation takes place in the listener's brain in the course of routine, nonlinguistic cognition as well. The distinction of language is that it is capable of instilling in the listener percepts, thoughts, and actions pertaining to situations that are not—indeed, cannot be—immediately present, such as tomorrow's weather or the conversation between Kalypso and Ulysses. Behavioral studies and brain imaging show that a person listening to a story gives in to a kind of guided hallucination, in which the representational processes unfolding in the mind's war room reflect not just the directly perceived world along with the internally generated anticipatory data and plans, but also an invented world that is induced by the narrative.[20]

The critical step in communication is to impart proper structure to the utterances that do the inducing. Intent is multifaceted (the technical term is multidimensional), it may have complex dynamics (including simulated events that play out over time), and the communication channel afforded by a string of words is narrow, all of which makes it difficult to pack into an utterance enough clues from which the listener's brain can form a reasonably well supported guess as to what the speaker meant to say. In charge of this process is a cognitive structure that serves as the grammar of language—a network of *constructions* or templates that manage the selection, inflection, and sequencing of words.

Constructions range in their complexity from indivisible morphemes (such as "post-") that can be prepended to a certain class of stems to denote a particular temporal relation, and words (such as "dog"), to larger structures in which some of the elements may vary from one instance to the next, such as "X range[ε/s] in [ε/its/their] Y from U to V," which is the template that I used to put together this very sentence. This last one is like a partially filled form that you may be required to complete in a doctor's office. Parts of this form, such as the words "range," "from," and "to," are prespecified. Others, denoted by symbols (X, Y, U, V), indicate "slots" where other constructions of particular types must be inserted. Yet others, which appear between brackets, indicate choice: one of the listed items (or none, if ε is listed as one of the options) must be chosen from each set.

In each case of communication, the grammar of the language that is being used needs to recruit out of its midst a squad of constructions that would represent faithfully enough the intent of the speaker, all the while evoking reliably enough a reasonable interpretation of it in the mind of the listener. The members of this ad-hoc team must then interact among themselves so as to generate approximately the right words in approximately the right sequence for the job. (Mistakes in this process are definitely tolerated, and even if an utterance comes out as a dud, another attempt at communication would often set things straight.)

To understand how this process works, we must remember three things about grammars. First, constructions that make up a grammar are codified patterns of language use. Second, those patterns may be combined—strung together or nested within one another—to form an intricately structured network of paths between words. Third, this network of constructions grows out of an evolutionarily older, but similarly structured, network of representations that shapes all behavior.[21]

A guarantor of the decent performance of this system is the constant evolutionary pressure on the memes represented by the constructions in question. Populations of constructions, along with the links that connect them to each other and to the rest of the mind, evolve to mediate between intent and interpretation; those that underperform fall into disuse, get disconnected from the rest of the network, and eventually disappear in a puff of neurotransmitters.

Returning to the example of the construction anchored by the word "range," we may note that within the network of grammar it is surrounded by a "halo" of possible paraphrases, some very probable, others less so. Distances within this halo—a measure of semantic slack, as it were—correspond to shades of meaning.

Depending on how much slack is allowed, the statement that dogs range in size from miniature to humongous can be paraphrased by saying that they vary in measurements between tiny and very large, or that they differ in dimensions between compact and huge, or, stretching it a bit, that they fluctuate in magnitude between small and big. The twinge of unease that one may or may not feel about this last paraphrase (depending on whether or not it evokes in one's mind an image of a pulsating dog) is a handy estimate of the subjective semantic distinction between "range" and "fluctuate."[22] The network of constructions, in which neighborhoods are defined by contextual similarities and semantic distinctions, serves as the representation space for language—a garden of forking production and interpretation paths.

To see how language works, then, ask yourself, with Ludwig Wittgenstein, "Do I understand this sentence?"

> If it were set down in isolation I should say, I don't know what it's about. But all the same I should know how this sentence might perhaps be used; I could myself invent a context for it. (A multitude of familiar paths lead off from these words in every direction.)

> . . . Phrased *like this*, emphasized like this, heard in this way, this sentence is the first of a series in which a transition is made to *these* sentences, pictures, actions. ((A multitude of familiar paths lead off from these words in every direction.))

> Every familiar word, in a book for example, actually carries an atmosphere with it in our minds, a "corona" of lightly indicated uses. —Just as if each figure in a painting were surrounded by delicate shadowy drawings of scenes, as it were in another dimension, and in them we saw the figures in different contexts.[23]

Seeing grammar in this light, we realize that the network that embodies it may be quite idiosyncratic and that meaning is therefore at least partially subjective. This discovery may feel like a letdown for someone who expects language to be perfectly universal and perfectly supportive of communication, but it is of course perfectly in line with the central role of experience in shaping the mind. It is shared experience, which rides on top of shared environment and biology, that promotes effective communication. Because you and I inhabit the same planet, because our brains and our bodies are alike, and to the extent that our minds grew out of similar experiences, you can talk to me and hope to be understood.[24]

Dependencies All the Way Down

To understand and be understood is the hope of every newcomer to a linguistic community (as those of us who ever moved to a strange new country always remember). For one category of new arrivals, however, learning language touches, at least initially, upon weightier issues than fitting in at school or improving one's chances of getting into college. This category consists of those citizens-in-the-making who face the task of learning a language, any language, for the very first time: newborn babies.

As creatures whose well-being depends (critically at first) on the goodwill of others, babies are eager to learn to interact with people and with their environment, and are good at it. At the same time, the memes that constitute language survive and flourish, along with other aspects of culture, insofar as they are good at being disseminated and learned. Our computational understanding of cognition allows us to state precisely what it means for learners of language to be good and for linguistic knowledge to be learnable. As in all learning, the common denominator here is statistical inference—the only way of gleaning knowledge from mere data.[25]

Because knowledge is a means for organizing data, a baby learner who does not yet *know* much about the world is running the danger of being overwhelmed by what I called in Chapter 4 the maelstrom of sensory information. Or so it would seem: because knowledge transforms both the recollection of experience from which it is distilled and the appreciation of subsequent experience, the very same situation *feels* different to a baby and to a grown-up, forcing scientists to resort to guessing when they try to imagine what it is like to be a baby. A particularly memorable guess of this kind has been put forward by William James: "The baby, assailed by eye, ear, nose, skin and entrails at once, feels it all as one great blooming, buzzing confusion."[26]

Even if James guessed right and babies are indeed confused by the information assault, they do not remain so for long, which suggests that they are well prepared to deal with their experiences. As before, to give co-evolution its due, we should credit not just the learner but also the data by noting that babies' experiences, in turn, are often enough structured so as to facilitate being dealt with. In the acquisition of language, this push-pull division of labor is readily apparent at every level and stage of the process.

Very high on any language learner's agenda is learning the names of things. It may be, for instance, that the baby's community refers to an object that looks like this 🐕 as собака. Such an object/name pairing would be very easy to learn if the name were always attached to the object, as a kind of bar code (acoustic, gestural, or, as in a grocery store, printed). Alas for the learner, a dog (in Russia or elsewhere) may appear on the scene without being named, and the word that stands for it may occur in a conversation on which the baby is eavesdropping without its object being present. Even when the name and the object happen to co-occur, they may do so within a wide window of space-time that they share with various other objects and sounds.

The right thing to do in this situation is to treat suspected pairings of words with objects as provisional hypotheses and to look out for statistical evidence that would allow principled arbitration among them. This strategy would be infeasible if all possible hypotheses were entertained. Good learners that they are, babies solve this problem by cutting a lot of corners: they only ever consider the very few hypotheses that are compatible with some very strong prior assumptions about what is and is not likely to happen in their world.[27]

Being good at learning does not mean being rash or foolhardy: such corner-cutting is fully endorsed by the Bayesian theory of statistical inference. (Like blind mole rats, babies are Bayesians without knowing it.) The Bayes Theorem, as you will remember, prescribes how the prior probability of a hypothesis should be modified by new data. The priors for word learning are themselves learned over evolutionary time, so that even an absolute novice learner can effectively rule out hypotheses that are incompatible with a few basic assumptions built into his or her brain.

One such assumption is that words name objects or events that "hang together" in space and time. This would rule out, among others, the hypothesis that собака refers to a "combination" object, such as dog with a slipper in its mouth. With co-evolution in mind, we may observe that this assumption is derived from certain properties of the world we live in and has to do with the data's contribution to its own learnability: statistically speaking, a word for "dog + slipper" would be hard to learn, as well as unable to compete with "dog" and "slipper." Another assumption is that a hitherto unfamiliar word spoken by a caregiver refers to the most salient object in the present scene. This would rule out the hypothesis that собака, uttered in a room with an excited golden retriever in it, refers to the batik on the wall.[28]

Mastering the names of objects gradually makes it easier for babies to learn complex constructions—the multi-word patterns of usage that are the entry ticket to any human society. At this level too, the learners rely on assumptions that are themselves learned over many generations and are incorporated into the developing mind's computational toolbox. The most general principle at work here is that of constrained reuse: learners expect that labels for objects, events, actions, attributes, and qualities, along with a smattering of "service" or function words such as "and" or "that," will appear and reappear in various combinations that conform to certain statistical patterns.

The learners' need to resort to statistics to find those patterns exerts selection pressure on the population of memes that give rise to the patterns (and thereby keep language well ordered and therefore usable). The co-evolutionary drive turns the learners' expectations into a self-fulfilling prophecy: the patterns that thrive under this regime are those that are statistics-friendly. Meme selection pressure also determines what those patterns actually look like. Because the elements of language—phonemes, syllables, words—are digital (so as to promote error-free replication), the patterns in sequences of elements are necessarily defined in terms of the *dependence* of some elements on others. Formally, this corresponds precisely to the concept of conditional probability—our old friend from earlier chapters.[29]

The simplest example of a dependence pattern is the idiom. For instance, the collocation *kick the bucket* is defined by the high probability of *bucket*, given that the preceding words are *kick the*. The pattern in this case is slightly more general: *kick the* ___ *bucket*, where the slot ___ may be left unfilled, or, with lower probability, may be occupied by the word *proverbial* (any other word there would destroy the idiomatic meaning of the expression).

Collocation with zero, one, or more slots in it is in fact the most general kind of linguistic pattern. As a formulaic construction, it allows the speaker to express intent in a manner that is ritualized, hence likely to be interpretable by the listener. The particulars of the message are conveyed by selecting a filler for each slot out of a set of possibilities that may be large or small but is always limited and is specific to the construction in question.

Stringing such constructions together sequentially and nesting them (that is, filling a slot with another construction that itself has slots) supports the expression of extended, hierarchically structured intents and messages, which can thus grow to be as complex as any conceivable structure in the mind. The resulting system nevertheless remains manageable because it relies on a single organizational principle: the codification of conditional probabilities of some elements, given others. This same principle extends to the very fundamentals of the use of language for communication, insofar as it captures the dependence of the choice of constructions on the speaker's intent and on the situational context in which communication is taking place.

I find it extremely pleasing that a communication system that has evolved to fit inside a brain made of neurons boils down to the one computational operation that neurons implement most naturally—making certain represented quantities depend on others. Given that the system of constructions (unlike, say, the physiology of a liver, but very much like the actual connection strengths among neurons in a brain) needs to be learned from a corpus of experience, it also makes great sense that the slotted collocation construction can be learned by tallying conditional probabilities over the discrete elements of language.

To do so, the learner that monitors a stream of utterances needs merely to align and compare them. The parts that match across multiple utterances signify a recurring sequence or collocation, whose statistical significance can then be estimated by assessing the probability of the recurrence arising by chance. At the same time, a localized mismatch signals a slot in the collocation, the differing elements being the options among which a choice must be made.

If, like me, you are not fluent in the language of Homer, you can appreciate the nature of the computational opportunities for (if not the actual experience of) discovering structure in a profoundly foreign stream of data by examining the following passage. These are the six lines from *The Odyssey* whose translation appears earlier in this chapter ("My lady goddess . . . "). Here I offer them in the original ancient Greek, complete with polytonic accents, but with spaces omitted, to simulate the feel of regular speech, in which there are no pauses between words:

πότναθεάμήμοιτόδεχώεοοῖδακαιαὐτὸς
πάνταμάλοὔνεκασειοπερίφρωνπηνελόπεια
εἶδοςἀκιδνοτέρηημέγεθόςτ᾽εἰαάνταιδέσθαι
ἠμὲνγὰρβροτόςἐστισὺδ᾽ἀθάνατοςκαιἀγήρως
ἀλλὰκαιῶςἐθέλωκαιἐέλδομαιἤματαπάντα
οἴκαδέτ᾽ἐλθέμεναικαινόστιμονἦμαρίδέσθαι

By approximating the acoustic speech stream, which is analog (in the sense of Chapter 2), with text, which is digital, this example focuses on how a learner can bootstrap from knowledge of the basic categorical elements of language (phonemes, represented here by letters) to the discovery of words (statistically prominent sequences of phonemes) and other, more complex constructions. As noted earlier, all it takes is for the learner to align and compare the stream of data to shifted versions of itself. This task requires memory for temporarily holding data and a mechanism for matching the held data to incoming sequences, both very straightforward functions that can be performed by rather simple neural circuits.

If you scan the Greek text for places where it matches shifted versions of itself, you will indeed discover sequences of characters that appear more than once. One such sequence is καὶ, which happens to be the Greek for "and" (see if you can find any others). Of course, a baby learning Greek has no access to such privileged information and so must rely on a statistical significance test to decide whether to admit καὶ into its lexicon or discard it as a fluke.

A partial rather than perfect match between two sequences that is nevertheless statistically reliable signifies a construction with a slot, in which some variation is to be expected. In the present example, the last two lines end, respectively, with the sequences ἤματαπάντα and ἠμαρἰδέσθαι. Guessing that πάντα and ἰδέσθαι are words in their own right because they appear elsewhere in the passage (indeed, these are the words for "all" and "to see"), we are led to conclude that ἤματα matches ἦμαρ. A Greek scholar will tell you that these strings are in fact related by inflection: they are the plural and the singular, respectively, of the word for "day."[30]

The computational scheme for learning language that I have outlined over the last few pages provides for all the necessary functional components of this monumental task. Controlled experiments in language acquisition show that babies indeed solve it by leveraging a few strategic assumptions about the nature of the data they face and by applying statistical inference to the problems of reference determination and structure discovery.[31] The recipe for learning language seems to boil down to this, then: scan the speech you hear for recurring structures while monitoring its situational context, note statistically significant matches within the speech stream as well as between speech and the outside world, feed what you already know back into the discovery machine, and soon enough it will not be Greek to you anymore.

A great thing about this scheme is that it is perfectly suitable for complete novices: you don't have to know anything specific about the language you are immersed in before you start. Beyond that, it only gets better: the farther along you are, and the more words and other constructions you already know, the more good candidates for further analysis suggest themselves. Knowing more also helps you make use of context (which with time feels increasingly less opaque) in figuring out difficult passages: you start by learning what собака stands for, and in just a few years, overhearing a conversation at a party, you surprise yourself by guessing right the meaning of *je ne sais quoi*.

It Takes a Village

Before you start humming *All by myself*, though, consider this: given that language is inherently and fundamentally a social game, is "going it alone" really the best way to learn to play it? Because language is an evolutionary game played by many players, various benefits—both for the memes and for their human hosts—may ensue if listeners and speakers collude to make learning easier and if the structures they trade make collusion worthwhile. Language learning on all levels is indeed boosted by social interactions.

In learning names of objects, for example, the baby's assumption that a new word labels the most salient object would be more effective if a certain degree of cooperation on the part of the caregiver (which need not be conscious and deliberate) can be relied upon. Indeed, when speaking to a baby, people usually signal that they do so by imparting a special modulation to their voice; they also engage the baby's attention and draw it (by shifting the gaze) to the object being named.

On their part, babies are far from being passive receptacles of information: they often take the initiative in seeking knowledge about the game they find themselves in. Although babies can learn language by simply being around adults who communicate among themselves, participation in social interactions with adults and peers makes a huge difference. In learning names for things, in particular, the baby's retention of a verbal label works best if the caregiver offers it in response to the baby's own vocalization directed toward the object, and if it is delivered within an appropriately short time window.

Speakers also often make it easier for listener-learners to match utterances while seeking after the patterns of constructions and their usage. The basic operations of alignment, comparison, and statistical testing are made more effective by the tendency of successive utterances in natural speech to be variations on a common structural theme. The resulting variation sets—runs of structurally related utterances—are particularly common in child-directed speech, where about one-quarter of all sentences appear within one variation set or another. As an example, here is a two-phrase variation set, taken from an Italian mother's conversation with her very young child: *Dove sono; dove sono i coniglietti* ("Where are; where are the bunnies"). In contrast to the passage from *The Odyssey*, the matching parts here are very close in time to one another, making the corresponding cue to structure much more prominent.[32]

While caregivers are generally unaware that they produce variation sets, this and other social behaviors that help children learn do depend on the personality and mood of the speaker. The effects of reduced cognitive stimulation and social interaction—conditions that are often brought about by poverty or parental depression—are far-reaching and permanent. Plopping children in front of a TV will not prevent them from learning to speak, but these children will not master language as well as their socially engaged peers. Unhappiness thus has a way of perpetuating itself.[33]

As children become better speakers themselves, they gradually increase their contribution to dialogue, and with it their participation in generating variation sets. The prevalence in adult conversation of variation sets in which all participants have a say suggests that there is more to this phenomenon than helping children learn language. The striking scope of coordination between conversants is illustrated by this transcript of a snippet of kitchen conversation between two adults, which has been aligned to reveal matching parts (omitting line 2, which consisted of a 0.2-second pause, and line 5, where Lenore sneezed):[34]

		A	B	C	D	E	F	G	H	I	J	K	L	M
1	LENORE	so	your	mother			's		**happy**	now				
3	JOANNE		my	mother			's	**never**	happy					
4			my	mother	would	n't	be		happy	if	everything	was	g–	—
6											everything	was	**great**	
7										and	everything	**is**	great	

Corpus studies show that this kind of coordination is a rule rather than an exception in naturalistic dialogue, while functional imaging reveals corresponding coordination between the brain dynamics of speakers and listeners.[35] The labyrinth of nested constructions that embodies the knowledge of language is thus best thought of not as confined to a single brain but as spanning entire communities, with space always available for new additions to plug themselves in.

At any given time, the person who happens to be speaking chooses one among the many forking paths that run through the labyrinth. At each fork, the speaker's choice arises from a tug-of-war among the probabilistically weighted available options. The utterance that takes shape is thus conditioned on a variety of factors: the speaker's memories (experiential history) and brain dynamics (intent); the constraints imposed by the structure of the labyrinth (grammar); the environmental context within which the conversation is situated; and last but not least, the dynamics of other brains that participate in the exchange, which affect the speaker's brain most significantly through the medium of language.

By putting in place a general-purpose medium of communication, the memes that comprise language support the emergence and propagation of other memes. These include anecdotes of hunting and gathering, epic poems, pieces of gossip, reports of miracles, propaganda, news, slander, jokes, conspiracy theories, election promises, family and tribal history—in short, all those aspects of human culture that can spread by word of mouth.

The medium of language is symbolic and self-enriching. By distilling memories of use and context into shared patterns of structure and meaning, language empowers those who speak and understand it to build virtual bridges between brains and thus to trade mind states through what must seem almost like telepathy to the uninitiated. A passage brimming with meaning is, for those who can read it, like the friendly alien mother ship in *Close Encounters of the Third Kind*—a humble structure suspended in midair next to Devil's Tower, slowly revealing to the stunned humans as it turns upside down against the starry Wyoming sky a vast, scintillating constellation of light and sound.

Because language has what seems to be a direct line to the mind's emotional core, it can be deadly—but so can all the tools that our species ever invented, starting with the flint blade. Best learned with a little help from our friends, language is the most advanced contrivance we have for fashioning virtual worlds and for bringing our friends there. As such, language is an accomplished and exquisite tool for generating happiness—and, of course, for sharing it with the rest of the village.

❧ SYNOPSIS

What had been the boringly biological evolution of our species turned into a wild ride when some of our ancestors, habitually engaged in a constant cognitive arms race against all and sundry, got help from unexpected quarters—a bunch of behavior-turned-replicator memory patterns, or memes. The subsequent co-evolution of humans and their culture has been, and still is, sustained by two key properties of the loose coalition of memes that constitutes language: the categorical or digital nature of phonemes and words, and the constrained manner in which these elements can be composed to form hierarchical constructions.

Such constructions reuse familiar building blocks, yet allow for the expression of complex and potentially novel meanings in the standardized, hence interpretable, form of a sequence of sounds, gestures, or marks in a visual medium. These, in turn, serve as the hints and clues that the brain of a listener or a reader works from in its search for a reasonably well supported guess as to the speaker's or writer's intent. The constraints within the grammar of constructions that emerges through hierarchical composition take the form of statistical dependencies. Thus, words are highly probable sequential patterns of phonemes, while larger constructions are likewise defined by some words depending on the choice of others.

Its reliance on categorical representations and on statistical dependencies reveals the organic connection between language and the rest of cognition. (Insofar as every neuron's output expresses a pattern of conditional probabilities over its inputs, dependency is the universal calculus of computation in the brain, whose utmost goal is to support forethought by mirroring patterns of conditional probabilities linking prior experience to possible futures.) And yet, language is unique among cognitive functions in the degree to which—in the best co-evolutionary tradition—it both helps and is helped by social interactions. As social animals, we revel in group play, which is what language evolves to promote and we evolve to master. Happiness and misery being the two-pronged stimulus with which evolution prods its pack animals, is it any surprise that we can be moved to tears or to laughter by a few well aimed words?

6 | Nobody, at Home

The web of cause and effect. Through a scanner, darkly. Because it's there. Connecting the dots. Flow. Soul music. Being and time and zombies. That which we are.

You who are on the road
Must have a code that you can live by
And so become yourself

  —CROSBY, STILLS, NASH, AND YOUNG,
   "Teach Your Children" (1970)

Whoever sees the web of cause and effect, sees the Way.

  —THE PĀLI CANON,
   The Great Elephant Footprint Sutra
   (committed to writing in 29 B.C.E.)

Happy is he who was able to know the causes of things.

  —VIRGIL, *Georgics* (29 B.C.E., book II, line 490)

The Web of Cause and Effect

The computational chores that babies face in mapping the labyrinth of language in the middle of which they find themselves upon arrival are closely related to the broader cognitive task of learning the ways of the world, of which language, as Borges would put it, is an incomplete, but not false, image. To see how well language fits within the rest of cognition, ask yourself, "Where is this paragraph going and what does it mean?" and then substitute "situation" for "paragraph." (This far into the book, I bet you have a reasonably good idea where my paragraphs go, given how well you have mastered the science of reading my mind.)

Both in language and in cognition in general, mastery comes down to the same two abilities: first, *understanding* the world by seeking patterns in sensorimotor activity and learning to relate them to a wider context, including your own and other people's experiences and mind processes; and second, using understanding to support *foresight*. The big picture is in fact even simpler than that: understanding and foresight are really two sides of the same coin, because they both hinge on knowledge of the causal structure of the world.

In language, this knowledge allows a person to process an utterance so as to yield some clues regarding the state of affairs of the world, including the utterer's beliefs and intentions, and, in due course, to process one's own intentions so as to yield a plan for generating an intelligible sequence of vocal and/or other gestures. The mechanisms of language learning keep tabs on experience, seeking to distill from it causal patterns of dependencies, such as the effects of discourse and situational context on construction choice and sequencing. These mechanisms build on more general cognitive capabilities that seek patterns in episodic memories. The knowledge of language, or grammar, is thus part of a wider web of cause and effect linking people and situations, and so ultimately is a part of the code of conduct of the world at large.[1]

Through a Scanner, Darkly

To yield meaning, the raw stream of ongoing experience, along with episodic memories of the past, must be searched for statistically reliable patterns. Like ripples on the surface of a pond haunted by a school of koi, these patterns arise from objects that are immersed in the world; a particularly perceptive frog sitting on a lily pad and watching the water surface could perhaps infer the koi's presence and learn about their movements.[2] Some such patterns are relational in that they depend less on what the objects are and more on what events they participate in and on how these unfold in time: to a koi watching from below, the pattern created by a pebble skimmed off the water surface would appear very similar to the ripples produced by a skillfully aimed small frog.

Once learned, patterns that are characteristic of objects and events become the nodes and the links in the mind's representation of the world-wide web of cause and effect. Far from being a mere record of experience, this representation brings out the regularities in it, which alone afford understanding and foresight. Whereas any singular experience, in every respect unlike anything you ever encountered, is an absolute mystery, two of a kind are a revelation that may help you prepare for a third. Having evolved in an unforgiving world that often enough is also just so partially predictable, we are geared toward perceiving and thinking in terms of generalizable patterns of information, or *concepts*.[3]

It is frustratingly difficult to explain the concept of "concept" by offering and discussing examples—not because examples of concepts are scarce, but rather because concepts are an integral part both of the process of explanation and of its intended product, comprehension. This quandary stretches from the loftiest realms of abstract cognition, where one hopes against hope that language-mediated conceptual inquiry can resolve conundra of its own making—such as "What is truth?"—to the most concrete concepts that seem to be directly and immediately grasped by the senses, such as a straight line or an ascending tonal sequence.

I figure that the concept-ladenness of abstractions such as "truth" (let alone "concept") is pretty obvious as it is, so let us focus on concrete stuff—say, a thin straight line traced with a sharp pencil on a wide expanse of white paper. The hard fact of the matter is this: nowhere in the brain's visual pathways is the representation of this line thin, or straight. To begin with, its projection onto the retina, which conforms to the inner surface of the eyeball, is curved. This arc-like retinal image is sampled by an array of photoreceptors, which are tightly packed into a mosaic that is locally roughly hexagonal, but otherwise irregular. Within this array, the set of receptors that actually respond to the line—more vigorously, or less so, depending on how close the receptor is to the center of the line's retinal "footprint"—is shaped like a banana.

A couple of intra-retinal processing stages later, the eye's million-dimensional output representation space is defined by the axons that form the optic nerve, one dimension per axon. Applying the no. 4 conceptual tool from Chapter 3, we recall that in this space the line is represented by a single point. The computational problem faced by the rest of the brain—to determine, without outside help, that this point in the representation space stands for a straight line "out there" in the visible world—may not be very difficult, but it is not trivial either, and its very existence suggests that perception, even of the simplest imaginable stimuli, is never simple or direct.[4]

The collective, distributed computations through which visual concepts are related to the outside world need to be calibrated and maintained. The brain does it by continually adding lessons from its ongoing experience to the pool of statistical knowledge that embodies each concept. Statistically speaking, a straight line is the average of a population of lines that curve in opposite directions. Evidence that the brain maintains this average dynamically and uses it to calibrate its perception of straightness comes from studies in which people are asked to wear prismatic goggles that distort the perceived environment by causing straight edges to appear curved to one side. After a period of adaptation, the subjects cease perceiving the curvature, at which point the goggles are removed and the opposite effect emerges: they now see straight edges as curving to the opposite side. This and many other findings show that perception is *situated*—rather than dealing in universals, it is attuned to certain collective properties of the particulars of the perceiver's environment.[5]

In its constant strain to stay calibrated, the brain responds not only to the statistics of perceived external stimuli but also to the many internal status indicators generated by the body that it helps to control. Thus, hikers facing a hill see it as quantifiably and significantly steeper if they are carrying heavy backpacks or are hungry, compared to the perceptions of unencumbered subjects who just finished a nice breakfast. When examined in this manner, the mind is revealed to be thoroughly *embodied*. Even the highest-level cognitive functions are affected, as are emotions—processes that compute and broadcast cues that modulate the rest of cognition. Nod your head as in agreement while listening to a sales pitch and your attitude toward its message will be more positive than if you shake your head as in disagreement; hold a pencil sideways between your teeth so as to stretch your face into a semblance of a smile, and you'll find a joke that's being told to you funnier than if you hold it in your mouth as you would a cigar. It seems that any conceivable claim regarding how bodily states may affect mind states turns out to have some substance to it, if only one cares to investigate.[6]

The two complementary principles of cognition that I just sketched, situatedness and embodiment, demonstrate how the virtual computational construct that is the human mind is rooted in the human brain, body, and environment. This new understanding elaborates upon the by now familiar notion that the world we perceive is virtual by explicating the nature of concepts—the computational processes and representations that give rise to our perception of reality. As expected from a complex statistical computation, this perception may or may not be veridical, depending on the task and the situation.[7] Because of the unavoidable uncertainty and the occasional outright distortions (some going unnoticed, others revealed as perceptual illusions), conceptual structures are sometimes described metaphorically as a prism through which one perceives reality. We now know enough about the nature of minds and their embodiment to realize that this metaphor is grossly misleading. There is no functional counterpart in the brain to a looking glass or a scanner that is separate from the self that sees through it, just as there is no divide between the self and its past interactions with the world. Had I been born with a different kind of perceptual system in place, or had my subsequent experiences been radically different, I would have been a different person now.

Because It's There

Many of the representations at the mind's thinking "core" are active processes that are constantly on the lookout for grist for their computational mills: perceived environmental affordances; episodic memories; bits and pieces of presently applicable more or less general knowledge; inner motives and drives; and whatnot. The products of their computations are turned into more memories and more knowledge and sometimes bring about overt action. Because actions have consequences, this thinking core strives at all times to perceive the world (and itself, when it turns its gaze inward) not as a random assemblage of informational odds and ends but as a meaningful web of cause and effect. Because of that, and to distinguish it from other components of what I am, I shall call this part of my mind the *effective Self*.

Although it goes a long way toward defining my unique personality, the effective Self is not confined to my skull, for the simple reason that the concepts represented internally within my brain are hopelessly entangled, in the cause-and-effect sense, with outside objects and events. Because external objects are the causes behind their internal conceptual representations (mediated by the clusters of sensory and relational features that the concepts account for), they are poised to contribute significantly to the determination of behavior. Speaking for myself, I can observe, for instance, that a significant (albeit by no means exclusive) determinant of my behavior is a geological formation in faraway California: Death Valley.

I had always been strongly attracted to the peace and solitude of desert landscapes, but assorted obligations and the logistics of getting to a desert from the all too civilized parts where I live used to keep me from hiking as much as I would have liked to. Finally, a few winters ago, I decided that the following March I would set aside all other plans for a week, fly to Death Valley, and go hiking. Since then, several times every year I disappear into one of the great deserts of the Southwest for a few days at a time.

Don't be misled by the conventional phrasing of my description of the decision process in the preceding paragraph ("I decided . . . ").

In truth, the "I" who made the decision includes a chunk of California wilderness. Were it not there, the world would have been a noticeably duller place, and my plans for the following March would have been different. (This observation singles out a sense in which George Mallory's stated reason for climbing Mt. Everest—"because it's there"—is literally true.)

On a less far-out but equally important note, my effective Self also maintains a bidirectional (albeit far from symmetrical) causal interconnection with the university that is my present employer: were it not for the spring break that it offers its students, in March I would be teaching in Ithaca, New York, not hiking around Panamint Springs, California. And, closer to home, a far stronger connection links my own Self with that of my wife, who cheerfully tolerates my wilderness walkabouts, if only because causation in this case runs both ways: I am a part of her just as she is a part of me.

Catching up with certain long-standing poetical views of the nature of the Self, cognitive psychology is opening up more and more to the idea of the intermingling of selves as an explanatory device both for falling in love and for staying there. On this account, the melding of the selves of long-wed spouses is merely an extreme case of the common crossing-over of trajectories of cause and effect between the brains of people who are open to each other. This latest news from the psychology of love would have pleased John Donne, in one of whose epithalamions we find this line: "You, and your other you, meet there anon."[8]

Connecting the Dots

The brain's compulsion to seek patterns of causation applies most strongly to observed outside data. It is undaunted by the

fundamental lack of certainty in inferring causation from correlation, settling instead for likelihood. On a moment-by-moment
basis, a strong illusion of causality ensues from uncanny coinci-
dences in which one event follows another in close succession. If
a mug sitting on the edge of my desk falls off and shatters as soon
as I point a wand at it and mumble a spell, you're bound to feel
that magic is afoot, even if rational inquiry reveals alternative ex-
planations (such as a trick with a piece of fishing line) that are
more plausible.

As you probably already guessed, the computational framework for making this kind of inferences is Bayesian probability.
Cognitive science now views "rational" as synonymous with
"Bayesian"; the rational explanation of the mug magic merely
does the right thing of taking into account the prior probabilities of the occurrence of magic (vanishingly low) versus sleight
of hand (non-negligible) in the sum total of human experience. In
a similar vein, our perception of causality is governed by the observed contingency between a suspected cause and an effect, defined as the difference between the conditional probability of the
effect given the cause and the conditional probability of the effect
in the absence of the cause.

If an opportunity arises for it, intervention—poking the mi-
rage with your finger, as it were—always trumps mere observation. If while hiking along a canyon rim you clap your hands and
there is a peal of thunder mixed in with the echo, you may get the
impression that you caused the thunder. Before concluding that
you have power over the elements, clap your hands a few more
times to see if thunder follows in the same close temporal succession; if instead it rumbles at random intervals, you're power-
less in the face of the impending electrical storm (and you better
get down from that exposed place quickly).[9]

Although canyons, universities, and spouses can all be parts of someone's effective Self, the person in whose brain reside the representations of such diverse factors is usually the entity that is held accountable for actions in which he or she is involved. (This, by the way, is why the claim "Death Valley made me do it" is not going to be an effective defense if you're charged with grand theft auto after being nabbed driving a stolen car down Emigrant Pass Road.) The shorthand for this highly effective heuristic view of consequentiality and responsibility is *agency*—a trait that people instinctively attribute to others and to themselves.[10] While being inseparable from the web of cause and effect, an agent comprises a sufficiently distinct neighborhood therein—a persistent causal nexus in which many chains of events come within striking distance of each other.

The attribution of agency to others is probably the most important kind of causal inference that you can make. Agents—a category that traditionally consisted of other animals and perhaps the odd carnivorous plant but must now be expanded to include artificial autonomous entities such as guided missiles—may be actively out to get you. In comparison, dumb nature's attempts on your life are always random (it may not be much of a consolation in case you get struck by lightning, but at least you know it's not personal).

It seems that in discerning agency among nature's random tricks, people tend to err on the side of caution; "such tricks," observes Theseus in Shakespeare's *A Midsummer Night's Dream*, "hath strong imagination":

> *That, if it would but apprehend some joy,*
> *It comprehends some bringer of that joy.*
> *Or in the night, imagining some fear,*
> *How easy is a bush supposed a bear!*[11]

Because the bar for inferring agency must be so low, the brain looks out for anything that seems out of the ordinary. An overly regular series of events may trigger an apprehension of agency: an acorn hitting me on the head as I sit under an oak tree I can write off to chance, but getting pelted by half a dozen acorns would start me wondering about scheming mutant squirrels. (And if I run through the grove in a zigzag and still get hit by a bunch of acorns, intentional malice on the part of some tree-dwelling entity is all but certain.) Unpredictable willfulness too is a sign of agency: an object that buzzes past my head on the village green may be a windup toy airplane, but if it changes course and approaches me from another direction, it is either radio-controlled, or (recall p. 82) a hummingbird.[12]

No less important than the detection of the handiwork of other agents is reliable attribution of agency to Self. This task is more urgent and less straightforward than it may seem. The Self is not a little tin soldier in the skull's cockpit, but a loose conglomerate of computational processes, some of which, moreover, extend into other people's heads and into the rest of the external environment. Given the distributed, multiple-process nature of the mind and the complex mechanics of the body, it is crucial for the ruling coalition of those processes to assert control and to ascertain that its control is effective. Unless the causal nexus that is the effective Self is properly labeled as such, no credit or blame can be properly assigned where they are due, and hence no learning from experience can take place. Furthermore, human social functioning depends on the availability of a "person" construct that supports the tracing of interactions and the administration of justice.[13]

As a clump of pathways of cause and effect that snake through the represented web of possibilities, the basic "person" construct is what we called a concept—a probabilistic model of a small part of the world—which coexists in the mind with other concepts, such as "straight line," "breakfast," and "marmot." These models are all generative: they can be made to produce likely instances of the objects or events they represent. These, in turn, are used to simulate and predict the disposition of the corresponding entities, including other agents, in a given context. The model of the Self is no different in this respect: to plan a course of action, I need merely to weigh and elaborate upon a variety of predictions generated by the computational processes of which I am composed.[14]

Learning to understand the workings of the world, of other people, and of our Selves by building probabilistic models and using them in simulations is not an easy way to make a living, and yet, as a generalist species selected for smarts, we humans do just that.[15] What is it that motivates us? Learning requires an expenditure of personal effort and resources, so one expects that good learners would be well motivated personally (over and above being selected for from one generation to the next). This intriguing thought suggests that there is a connection between effective cognition and happiness.

Flow

Those of us who served time as parents or educators of school-age children will attest to the glaring contrast between the ease with which most young students pick up everyday physical and social skills and the hard work that they must put in to master formal knowledge. An indifferent student who long ago embraced the notion that math is hard and who nevertheless wakes up in the morning with a newly acquired passion for calculus would be justified in seeking signs of a covert pharmacological intervention on the part of the usual suspects—the overzealous parents. In the acquisition of life skills, the usual suspect behind the general ease of learning is evolution, which, indeed, is guilty as charged.

Given the volume and the sophistication of computation that goes into physical and social cognition, it would be wrong to assume that we sail effortlessly through everyday learning because the tasks that it encompasses are intrinsically easier. Rather, learning the world feels easier to us because we belong to a species that evolved to be good at it and to feel good about being good. In particular, it feels good to be able to explore new places and meet new people, to discern a pattern in hitherto seemingly random events, to gain deeper understanding by deploying knowledge and honing skills, to anticipate developments while appreciating the unexpected . . . let's admit it, there is a lot of fun to be had in simply being able to learn new stuff.

The hypothesis that we are naturally predisposed to enjoy everyday learning is supported by two related sets of empirical findings. On the one hand, and perhaps not entirely surprisingly, effective acquisition and use of knowledge and skills contributes to positive affect. On the other hand, positive affect (which may be induced by such simple interventions as being offered a piece of candy before the experiment) promotes association of ideas, pattern detection, decision making, problem solving, and creativity.[16] This virtuous circle (a designation intended to highlight the self-reinforcing nature rather than any intrinsic moral value of the phenomenon in question) has all the trappings of an out-of-control evolutionary ploy for taking over the world—a project that our species is, by any account, pretty deep into by now.

The open-endedness of this project is virtually guaranteed by the hedonic advantage of the *process* of learning over its completion—of pursuit over accomplishment. In happiness research, the condition of ongoing enjoyment derived from being engaged in an activity that is challenging, but not frustratingly so, is called *flow*. The likelihood of a person experiencing flow can be estimated as the proportion of time spent above his or her mean level of challenge and mean level of skill. (Within this framework, being above one's mean skill but below mean challenge is relaxation; high challenge and low skill is anxiety, and low challenge and low skill is apathy.)[17]

It seems that flow can be experienced in association with any sufficiently demanding activity that is potentially enjoyable. For my part, this includes going on long solo hikes in the desert, preferably to places I haven't been to yet;[18] brainstorming with my students and colleagues; skiing slopes that are near the limit of my skills (but not pushing it—there is nothing like a broken leg to interrupt your flow, as I know from experience); swimming the last half-dozen pool lengths of my workout quota (when I no longer need to conserve strength and can go all out); and, on rare occasions, writing (which normally feels like hard labor, in the judicial-punitive sense).

Positive affect has measurable, significant benefits for decision making and problem solving. When experienced as flow, in a task that is meaningful, interesting, or important to the person, positive affect prompts decision making that is both more efficient and more thorough.[19] In problem solving, positive emotions make people more likely to switch from stereotyped behavior to novel, more open and creative approaches.[20] We can tell what this means in terms of the computations that constitute thinking: within the tangle of representations of cause and effect that form the effective Self, positive affect loosens the associations and deepens the look-ahead into possible futures. As a result, unusual and creative avenues of exploration become available to the processes that analyze the situation and plan ahead by simulating likely chains of events.[21]

Soul Music

Any stand-alone information processor that must fend for itself, be it a hamster or a human, has a functional need to distill its experience into an effective Self, but only humans supplement theirs with stories they compose and trade—about their social encounters, their exploits, successes and failures, their aspirations and plans. The growing, often rehearsed, incessantly revised personal anthology is the *narrative Self*.

The stories that make up the narrative Self are a conveniently packaged extension of the grammar that embodies one's knowledge of language and also one's knowledge of how the world works. In this second capacity, the narrative Self complements the nonverbal behavioral scripts that apply to various common situations. Among the great many scripts at my disposal are those that specify how to get from home to work, how to build a superheterodyne shortwave radio receiver, how to cook *mapo doufu* from scratch—and also how to ask a policeman for directions without getting arrested, how to describe the state of my health if asked, and how to tell my favorite Jewish tailor joke.

Many of these narratives codify autobiographical information—self-censored, and often enough tweaked, embellished, or even largely invented, not necessarily with malicious intent—that we deliver to others on various social occasions or recount to ourselves for self-maintenance purposes. Our narratives are complemented by the narratives of our friends and acquaintances in which we happen to figure more or less prominently and by snippets of information about us in external media.

Similarly to the parts of the effective Self being distributed across multiple brains and their physical environment, all of which deliver causal nudges that contribute to its decision making, the narrative Self extends out into the media that can harbor the right kinds of information. (These too can participate in steering behavior.) We know that insofar as minds are persistent, causally effectual information patterns, they can in principle be hosted in any medium with the right dynamics, such as appropriately structured electronic circuits. When we think of minds in terms of distributed effective and narrative Selves, we realize that, in a real and increasingly important sense, parts of our minds already are outsourced.

The first example of an electronically outsourced chunk of my Self that comes to mind is a Google search alert that I confess to having set on my full name a few years ago. I have absolutely no idea where this chunk resides: the physical locations at which copies of the relevant piece of search software and the results they return are hosted would be difficult to pinpoint without detailed knowledge of the architecture of the Google system and its moment-by-moment operation.[22] Despite this extreme geographical uncertainty, the search results may get incorporated into my narrative and effective Selves, as when a new review of something I wrote ends up having repercussions for my behavior (such as prompting me to make another resolution to stop reading reviews).

Extrapolating this state-of-the-art search technology into the future leads us to the idea of a software mind spinning off an autonomous avatar that is sent off into cyberspace to carry out tasks on behalf of its originator and that eventually returns to be reintegrated into it, along with the stories and experiences it brings back. This idea, which I recently encountered in a science fiction context, probably has its roots in Chapter 13 of the Book of Numbers, which describes how twelve men sent by Moses to spy out the land of Canaan return with their stories to their respective tribes, who have been waiting in the Paran wilderness. The repercussions of the stories told by the spies are dire: the news of Canaan, which all but two of the spies described as "a land that eateth up the inhabitants thereof," makes the people rebel against Moses and Aaron and clamor for a return to Egypt, for which they are condemned to wander in the desert for forty years.

The point of the story you are reading right now is not merely that stories can be causally potent but that they can become incorporated into Selves. This is a strong claim, for which a correspondingly strong example is in order, and it would be hard to find a better one than that of a Self who is *all* narrative: Ulysses. In Book XIII of *The Odyssey*, the Ithacan's own facility with language earns him divine praise. After he tells a particularly ingenious pack of lies about his purported origins to Athena, who appears to him in a mortal shape, she drops her disguise and pretends to chide him—

> *Whoever gets around you must be sharp*
> *and guileful as a snake; even a god*
> *might bow to you in ways of dissimulation.*
> *You! You chameleon!*
> *Bottomless bag of tricks!*

—only to note that he and she are two of a kind in that respect. No matter that this and other stories about Ulysses may be a far cry from the real, historical king of Ithaca; Homer's narrative makes him come alive and makes us treat him (automatically and instinctively, because that's what we're wired to do with stories) as a person—on par perhaps with some remote acquaintance or a long-dead relative, whose would-be thoughts about certain things we may be able to imagine and whose imagined behavior may influence our own.[23]

As befits a cognitive representation and a complex meme, a story that is part of the narrative Self is an active entity: not a page of sheet music, but a wandering musician peddling a tune—as Hume might have put it, a citizen of the republic of soul. The dynamics of the mind—the music of the soul—arises from many such entities acting in concert.[24] Its beginnings, in an embryonic nervous system that is just starting to resonate to the world outside itself, are modest; its potential for development and learning is immense. As one travels through life's "vale of soul-making," so poignantly perceived by Keats, the republic of soul grows populous and its music takes a distinctive form. To understand the functioning and the full representational range of its dynamics is to understand how and why *this* is what it feels like to be alive.

Being and Time and Zombies

Echoing the ancient wisdom according to which the proper road guide to happiness is self-knowledge, Paul Bloom recently observed: "We used to think that the hard part of the question 'How can I be happy?' had to do with nailing down the definition of *happy*. But it may have more to do with the definition of *I*."[25] Happily, instead of merely defining *I*, we can now do much more: we can *explain* it.

The explanation that has taken form in this chapter includes, but is not exhausted by, the executive and the narrative components of the Self—aggregates of computations over a web of conceptual representations of the world, ultimately rooted in statistical patterns of perception and action. The hitherto missing component is the *phenomenal* Self: what at this very moment—now!—it feels like to be me.

For a physical/computational system, there being something (rather than nothing) that *it is like to be it* happens to be a rare mark of distinction. For instance, it most probably doesn't feel like anything at all to be an anvil, whether it rests quietly on the ground or is busy computing its trajectory after being pushed off a castle wall by a ballistics officer. Likewise, intuition insists that being a marmot, or even an ant, must feel like something. Cognitive science is just beginning to understand, in computational terms, the nature of the scale of capacity for phenomenal experience, at the bottom of which are systems that are capable of none, such as rocks and plants, and on which an average human occupies some intermediate rung.

It seems that a system's capacity for experience is determined by the intrinsic distinctions that exist among categories of its own states—which, in a cognitive system, represent the states of the world of which it is part. The breaking-down of states into categories must be intrinsic to the system because its experience is. Not only is it up to me to perceive and feel the way I do: an examiner situated outside my brain would have to work very, very hard even just to establish through observation that I am conscious and am having phenomenal experiences, let alone figure out what they are.[26]

The identity between a system's ability to serve as a vehicle for phenomenal experience and the intrinsic categorical complexity of its state space implies that any experience is inherently extended in time. In other words, the moment-to-moment content of experience must consist of the system's *trajectory* through its state space. Freeze time, and all systems with the same number of distinct elements—neurons in a brain, grains of iron in an anvil, or whatever—become functionally equivalent in principle, insofar as their state spaces have the same dimensionality; let time roll, and only those systems whose intrinsic dynamics afford it—brains, but not anvils—proceed to have experiences.[27]

The categorical distinctions among the brain's possible trajectories through its state space arise from dynamical constraints inherent in the connections and prior states of its neurons. What it is that it feels like to be me is determined, therefore, moment by moment, by my entire nervous system's functional anatomy and physiology and by its interactions with the rest of the world (including, of course, the history of such interactions, which has contributed to shaping my effective Self). Phenomenal qualities that send this system on distinct trajectories get experienced as such, in any experiential domain.

For me, some such qualities are the apricot color of yesterday's sunset, as distinguished from the cranberry color of today's; the velvety explosion of taste of an Australian Shiraz from the Barossa Valley, as distinguished from a subdued French Médoc; and the feeling of satisfaction that comes at the end of a long bicycle ride, as distinguished from a long ski run. In comparison, on the level of phenomenal experience I cannot relate at all to qualities that my cognitive system, for one reason or another, conflates, such as the appearance of two St. John's wort flowers that differ only in the ultraviolet; or the taste of a Château Plagnac from 2005 and another one from 2006; or the feeling of opportunity evoked by two configurations of pieces on the game of Go board.

Seeing that improved discernment is a frequent (albeit by no means obligatory) side effect of accrued experience, we begin to understand how people's sensitivity to the so-called small things in life, both good and bad, and their capacity to enjoy those sensations that feel good can grow with age. Clearly, the richer my conceptual structures, the richer my phenomenal experience, and the closer it is to the ultimate reality—something that we can only know through our experience and that may therefore be usefully defined as the sum total of all possible ways to experience the world.[28]

But why is it, you may wonder, that the process of discernment—the state of the cognitive system ascribing a trajectory through a structured space of possible representations—should feel to it like anything at all? Why is it that the dynamics of discernment is what makes the system's perception of the world experiential? Such doubts about the identification of experience with the dynamics of intrinsically structured state spaces are understandable: because of how difficult it is to ascertain from the outside that a given system has any experiences at all (let alone experiences to which one can relate), philosophers customarily treat reductive phenomenology—explaining experience in less than mysterious terms—with extreme skepticism.

A much-used theoretical construct in philosophical phenomenology is a zombie, which by definition is a human being in all respects of composition and behavior, except one: it is incapable of experience. Because such beings are conceivable, the argument goes, all attempts to reduce experience to something that is not already obviously experiential, such as computational processes in the brain, are undead on arrival.

This line of reasoning is, however, without merit. The apparent conceptual possibility of something or other is not a sound foundation for understanding the world. For instance, my ability to imagine an electron without charge attests to my disregard of the corpus of knowledge generated by physics rather than to the epistemological status of any physical theory. This example is particularly apposite because, on the reductive account that is being offered here, experience is just as inherent and essential a characteristic of the dynamics of certain computational systems as charge is of the electron.[29]

Its vaunted conceptual possibility aside, a species consisting entirely of experience-lacking zombies would be at a serious evolutionary disadvantage. The members of a species that lacked all phenomenal experience could know no pain or pleasure and would therefore be indifferent to the environment's ongoing appraisal of their actions. Collectively, such a species could still learn (through selection by early demise, which is likely in the case of drastically underperforming individuals in the wild), but it could hardly compete with species that engage in collective lifelong learning from experience, guided by the felt consequences of individual actions.

That Which We Are

Whereas detecting phenomenal experience from outside the system that embodies it takes some hard work in psychophysics and reverse engineering, from the inside it feels most immediate (that is, in no need of mediation). To us humans (and for a number of functional-anatomical reasons, likely to all animals with the standard vertebrate body and brain plan), it also feels *personal*. Right now, *these* perceptual experiences and *these* thoughts about them are mine, and so are *these* fingers, which are moved by me over the keyboard of a notebook that also happens to be mine, but in a less immediately felt manner.

A careful examination of the personal qualities of the phenomenal Self reveals four functional ingredients: spatial perspective (I perceive myself to be *here*, right behind the bridge of my nose); ownership (these experiences belong to me); agency (I am the originator of my thoughts and actions); and selfhood (at the core of this throng of experiences is an experiencer that is me). This analysis is supported by studies of pathological cases in which some of those ingredients of the phenomenal Self malfunction and by psychological studies that temporarily induce such malfunctions in the laboratory. For instance, an out-of-body experience can be induced in perfectly normal subjects by having them view (through computer-connected goggles) the back of a virtual human being stroked with a stick, while their own back is stroked in synchrony with the virtual setup. This simple procedure causes the subjects to perceive themselves as located way out in front of their actual physical bodies.[30]

The possibility of subverting our sense of spatial perspective by means of virtual reality manipulation is very telling: it draws our attention to the undeniable fact that the reality we inhabit is virtual to begin with. Indeed, were it not, how could I be looking out at the world through what to my sense of sight seems to be an oval opening in the front of my face, while my sense of touch (as well as common sense) tells me that what's really there is my nose? The realization that the mind deals in virtualities has been with us for some time, but now it comes to a crunch: the mind's *I* is virtual too (hence the title of this chapter).

The computational processes that jointly amount to a simulation of the experiencer are quite straightforward. The spatial perspective is provided by a combination of directional optical sensing from two vantage points (the two eyes) and fusion of the two sensed images into an integrated, map-like representation that includes viewer-centered simulated depth information. The sense of ownership and agency stems from an ongoing comparison between perceptual outcomes predicted through simulation (what it would look and feel like if this arm is indeed mine and if I decide to move it thus) and actual ones (what it in fact looks and feels like). As to the core Self—the dweller at the virtual unitary vantage point, the owner of the body, and the puller of its strings—it is merely a useful illusion that conveniently ties together these three privileged streams of experience into a single causal knot.

The illusion of the phenomenal Self as the unitary experiencer, decider, and actor is useful for the very same evolutionary reason that phenomenal experiences are. It is this virtual entity that serves as a causal bottleneck for the world's responses to one's actions, taking the blame for the entire extended system's failures and credit for its successes, translating these into practical cues for improving one's running score, and thereby ultimately making learning from experience possible on the level of an individual and not just the species.

Insofar as the extended system—the effective and narrative Selves planted in the great web of cause and effect—can reach deep into geographical and social space, the computational need for such focusing of behavioral feedback is pressing. By taking responsibility for the processing and use of this information, the phenomenal Self gives rise to another useful illusion: that of free will. In the West, it has been known since at least the time of Voltaire and Hume that the concept of uncaused cause, which is a prerequisite for free will, is logically incoherent: if my "free" decision to do a particular deed arises absolutely independently of any of the existing circumstances, including my own prior actions and states of mind, then in no sense can it be considered free, or, indeed, mine.

Despite the hopelessly messed-up logic behind it, the illusion of free will is useful because along with the illusion of a unitary self it provides a handy accounting mechanism for social justice and thus facilitates cultural learning and other types of social cognition. As some Buddhist philosophers intuited long ago, the web of cause and effect being the sole reality implies that there can be freedom only to the extent that there is not a Self. When considered in a modern cognitive science context, this tenet is seen to bring out a deeper understanding of the freedom of will instead of negating it. I do not *obey* a code of conduct that is implicit in the web of cause and effect; rather, I *am* part of the web. The ultimate reach of this web is universal—and within the boundaries of physical law the universe is, of course, free.[31]

Our quest for self-knowledge has thus led to the startling, yet ineluctable conclusion that everything about the mind—from the representations it constructs, through the simulations it conducts to plan behavior, down to the subjectively free phenomenal Self that seems to be running the show—is virtual, in a concrete computational sense. Of course, to this virtual entity itself it all feels very real, which is the whole point of the mind's going to the trouble of simulating a self in the first place.[32] To me, having this much insight into such an exceedingly complex system whose well-being I care about so viscerally feels liberating. And yet, it seems to me at times also that this knowledge casts a diaphanous veil of unreality over the world, which makes me think of Everett Ruess, a young artist who in 1934 disappeared into the southern Utah wilderness and who wrote in one of his last letters home: "Often as I wander, there are dream-like tinges when life seems impossibly strange and unreal. I think it is, too, except that most people have so dulled their senses that they do not realize it."[33]

✍ SYNOPSIS _____

The mind learns the world of which it is part by seeking patterns of statistical dependencies among experiential particulars, starting with premises and subject to constraints that are defined by genetic endowment and epigenetic development. From these particulars, which may be represented fleetingly or stored in episodic memory, arise the generalities that populate conceptual memory. To be useful for understanding and prediction or for forethought (which is the overarching functional requirement of any mind), concepts encode distilled causal knowledge about the world.

A persistent cluster of such conceptual knowledge accrued by a mind becomes the effective Self, which is part of the larger web of cause and effect that constitutes the reality we face and governs behavior within the constraints imposed by that reality. Being both social and articulate, we use language to augment our effective Selves with narratives, which, on a different level, also encode specific episodes from our own and from other people's personal history, as well as general knowledge about how the world works. Just like the causal strands of the effective Self, the stories that make up the narrative Self extend out into the world beyond our skulls.

Learning the code that one must live by is hard work for which we, as creatures that are subject to evolutionary pressure, are rewarded with transient effort- and success-related happiness—not because we are entitled to it, but because creatures that are thus rewarded learn better and are less likely to go extinct. To be happy, or for that matter to have any kind of feeling toward a percept, a thought, or an action, a cognitive system must have a capacity for phenomenal experience—a capacity that is predicated on certain structural qualities of the representation-space trajectories that the system's state can follow. Insofar as feelings are included in the evolutionary causal loop, creatures like us, which take life personally by virtue of constructing and using phenomenal Selves that feel, have a competitive edge over zombies that by definition do not (which explains why the latter are such a rarity in real life).

The coalition of Selves that is the mind uses its collective causal knowledge of the world to exercise forethought and to plan its own behavior by conducting simulations of likely scenarios and weighing its options. Simulating the world, complete with gravity, fruit-laden apple trees, other people, and perhaps a snake slithering through the grass, sounds a lot like setting up a virtual reality environment. This concept, which used to be owned by computer graphics experts, is now familiar to every gamer, but even if you are a veteran computer game player, you may be in for a surprise. What your mind takes to be the rock-bottom reality of the kind that is supposed to exist outside the gaming console is virtual in exactly the same sense that the simulated world inside the game is.

As a quick case in point, try to reconcile in some other way the single, uninterrupted, wide-angle quality of the visual world that you see with the fact that you are looking at the world through two eyes. The wide-screen visual experience that we are used to is virtual: it is the product of intensive computation, whose purpose is to recover depth cues from the two slightly disparate arrays of data provided by the eyes; when this or other computations that feed the mind's virtual reality fail, illusions follow. The mind's virtualized representation of the world encompasses not just the world's perceptual qualities: as we learned earlier, decision-making scenarios and action plans, as well as entire bundles of processes that represent certain parts of other people's minds, are also simulated.

Having realized that, you're but one step away from the epitome of self-knowledge: understanding the true nature of your Self. The bottom line, then, is this: the Self, along with all of its perceived and remembered attributes—anything and everything that is included in the feeling of being you—is a product of the brain's virtual reality engine. This virtual Self is computed and put in charge of the situation purely for reasons of good governance, that is, efficient and purposeful control of the brain's life-support system—your body.

7 | An Irresistible Call to Depart

Prometheus goes on parole. Alexander meets Diogenes in Corinth. History is made in Bishopsgate. Peace is struck in the republic of soul. Ulysses leaves Ithaca again.

You but arrive at the city to which you were destin'd—
   you hardly settle yourself to satisfaction, before you
   are call'd by an irresistible call to depart.
   —WALT WHITMAN,
      Leaves of Grass: Song of the Open Road (1892, 82:11)

Prometheus Goes on Parole

It is a beautiful, sunny June day here in Ithaca, in the great state of New York. On a fallen log in my backyard, a pileated woodpecker is having his lunch. A fawn whose mom parked it for the afternoon in the bushes behind the deck is napping. An overweight rabbit pretends to ignore the neighbor's sheepdog, while munching down a dandelion leaf, the narrow end first. Squirrels are pursuing each other all over the stand of white pine and look momentarily smug whenever they catch up with themselves.

On days like this, I am always moved to see the pursuit of happiness, which the United States Declaration of Independence lists as one of my inalienable rights, as just the thing for me. After all, I already have my life (good for as long as it lasts) and liberty (as much of it as the physics of the universe and my place in society allow). I have even experienced happiness occasionally (as recently as today), so I am sure I'll know it when I catch it again. But will I know what to do with it?

It would be nice if happiness, once caught, could be saved for later enjoyment, just as I preserve high-resolution digital photos of the idyllic—to the point of embarrassment—wildlife activity in my backyard, so that I can put them on my screen and remember June in late November, when the outdoor scenery in upstate New York brings to mind *The Seventh Seal* rather than *Bambi*. Alas, storage does not quite work with happiness. Even if the entire situation, not just a few megapixels' worth of it, were meticulously reconstructed, chances are it would not feel as happy as it once did. I can tell, because I have been there, more than once.

Let me illustrate what I mean with an example. I have been trying to understand for years why, given my love of hiking to wild and remote places, it is so hard for me to relax at the hike's destination for any significant length of time. Here I am, having scrambled up a deep desert gorge to the rarity of rarities: a flowing spring. It is barely past noon (I have been hiking since dawn), and I know that I have hours to spare before I must start on my way back to civilization, and yet I feel a growing restlessness. A chunk of cured sausage wrapped in a flour tortilla and accompanied by a pickle (good for the salt balance when hiking in dry heat) offers a welcome diversion, but fifteen minutes later the urge to depart becomes irresistible. I get up and walk on.

I used to think it was just me, but no, it turns out that human nature is responsible—more precisely, my position along the spectrum of restlessness, with which the members of my species are endowed to different degrees. It seems plausible that many of us would be perfectly happy to spend a long afternoon listening to the tiny desert brook that is the only source of water within a two days' walk, but who are those people? I suspect that they are the very same ones who are less susceptible to the call of the wild to begin with. If you would rather stay at home than go off on a walkabout at the first opportunity, you are more likely to find peace at any temporary stop along the way. The urge that drives people out into the wilderness is one and the same as the urge to move on that makes them restless once they are there.

This "call to depart," which may be likened to a chronic case of cabin fever, is just one particularly noticeable side effect of the gift of Prometheus: forethought. We know that animal species evolve the capacity for forethought because anticipating an imminent danger and generally thinking ahead promotes reproductive fitness. In humans, the capacity for forethought is particularly well developed. In our minds, we can revisit, reevaluate, and learn from past experiences, imagine and learn from events that could have transpired but did not, and simulate alternative futures. What is common to all such situationally creative thinking, which I described in an earlier chapter, is that it temporarily insulates the thinker from the surrounding reality. While savoring the memories of last year's vacation, wondering what kept you from getting out of the stock market before it crashed, polishing the phrasing of your next pay raise request, or even just trying to recall what that earlier chapter was about, you are effectively absent from the here and now.

Letting the mind wander in this manner is what the human brain does by default throughout waking life, as well as when dreaming while asleep. Some people like to wander the real world in person (this category includes myself and, posthumously, Walt Whitman), but even those who like to stay at home travel in virtual worlds whenever the cognitive demands of the present relent momentarily from occupying their minds. More often than not, default-mode simulated travel in space and time happens without conscious effort or even awareness of having strayed from reality. It does not matter that the mind's propensity to wander evolved in the first place because it happens to help with planning a course of action—now that it can part with reality and venture into virtual spaces, the mind is habitually on a hair trigger to do so. In this sense, we are doomed to be perpetually restless in the present because the mind is an embodiment of the anticipation of the future.

In David Cronenberg's expertly unnerving film *eXistenZ*, the characters play an immersion-style virtual-reality game, which they are given to believe they can suspend by exclaiming: "eXistenZ is paused!" In comparison, the reality that we are given to believe we inhabit can be suspended simply by thinking about something else. This insight greatly livens up Prometheus's gift of forethought, which otherwise would seem terminally boring. Prudence is good for you (as evolution and our parents never tire of reminding us), but it does not sound like much fun. What joy it is to discover that the same mental faculties that make you wise by helping you see that which is about to be born also free you from the tyranny of the here and now!

For stealing his fire and for endowing mortal creatures with forethought, Zeus chained Prometheus to a rock in the Caucasus Mountains, sending an eagle every day to gnaw on the Titan's liver (in clear violation of the Geneva Conventions). In romantic imagination, right is destined to win over might, and so in Shelley's *Prometheus Unbound* the tyranny of Zeus is eventually overthrown. It is heartening to observe that in this respect life imitates art, up to a point. The mind is free to roam, even as it keeps getting yanked back to reality every time the brain's physical well-being (as in evading the proverbial saber-toothed tiger) requires its presence. Prometheus's sentence is not annulled; he is out on parole.[1] All the same, having to check in every now and then with the parole officer of reality is not too steep a price for a mind to pay for otherwise unlimited access to a virtual world over which it reigns.

Alexander Meets Diogenes in Corinth

Given that each one of us must time-share between the real world and a virtual one, we might as well make the most of the situation. For that purpose, both the real and the virtual aspects of the situation matter: happiness, it stands to reason, depends both on one's station in the outer world and on the state of one's inner self. The implications of this straightforward existential insight are quite far-reaching, sometimes literally so: it has the potential of setting seekers after happiness off on diametrically opposed courses of action. When taken to the extreme, these opposites can be described as world conquest versus self-conquest. Let's take these ideas at face value and see where they lead us.

A would-be world-conqueror seeks happiness in increased control over his or her worldly circumstances. In classical times, one of the top role models for this kind of behavior was Alexander the Great of Macedon, who had been tutored by Aristotle and who managed to conquer most of what the ancient Greeks called the known universe before keeling over at an age when many people I know are barely out of graduate school. Alexander's short life may or may not have been a happy one: what he lacked in the way of personal virtue taught by his mentor Aristotle, he made up for in conquests.

In modern times, the need for physical conquest as a means of control has largely abated, because virtually any amount of control can be exercised through money. Even if it can't buy you love, money can buy anything else that you may care to try to make yourself happier. This fact of modern life underlies an observation that the widely syndicated contemporary philosopher Dave Barry attributes to his mother, to wit, that it is better to be rich and healthy than poor and sick.[2] If you manage to get rich without blowing your health, you've got it made.

The crucial question here is how rich is rich enough. More folk wisdom to the rescue: as recorded in the Mishna, the scholar Shimon Ben Zoma anticipated Dave Barry's mother by almost two thousand years when he preached contentment: "Whosoever is rich? He who is happy with his share."[3] Laozi's *Dao De Jing*, in the chapter titled "Curbing Desire," concurs: "The satisfaction of contentment is an everlasting competence."[4] The association of virtue with the curbing of one's desires has been contemplated even in the midst of a culture hell-bent on world domination: Diogenes of Sinope, called the Cynic, who was Alexander of Macedon's compatriot and older contemporary, professed to be content with having no possessions at all.[5]

Promulgating happiness by merely pointing to virtue and hoping for the best is hardly an effective way of helping others, even if the sage leads the way with a personal example. One wonders whether the good citizens watching Diogenes at his antics in the agora have ever been bothered by the suspicion that what really made him happy was how good he was at provoking them. Indeed, Diogenes could not sway to his views even Alexander, who must have been smarter than the average agora-goer (unless Aristotle graded on a curve). When the young king of Macedon and leader of all Greece, on his way to conquer the world, met Diogenes in Corinth, he was reportedly much taken with the old Cynic, yet his words of admiration were qualified: "If I were not Alexander, then I should wish to be Diogenes." As it were, they both remained themselves until the end, which, according to Plutarch, they met on the same day—Diogenes, at age eighty-nine, in Corinth; Alexander, just short of thirty-three, in the old palace of Nebuchadnezzar in Babylon.

The average sage's point-and-hope-for-the-best tactic in teaching the way to happiness through virtue reminds me of a Russian slapstick comedy from the 1960s, *Captive in the Caucasus* (no connection to the myth of Prometheus, as far as I can tell), in which a memorable toast is proposed at a dinner party: "I have a desire to buy a house, but I have no means. I have the means to buy a goat, but . . . no desire. Let us therefore drink to a future in which our desires always coincide with our means."[6] Unfortunately, as is the case with all manner of exemplary behavior, getting our runaway desires to turn themselves in at the precinct is easier said than done. To convince yourself that you are in fact content with something that you spurned the first several dozen times around you must seriously warp your value system. That, in turn, seems to imply that you must become a self-conqueror.

History Is Made in Bishopsgate

The volume of unsolicited advice one is liable to get when the word hits the street that self-conquest is being contemplated is staggering. Memes affiliated with all the religions you ever heard of turn up at your front door, often in pairs, jostling for porch space with pushy patents for self-improvement, which stand out because of their loud dust jackets. A scantily clad meme of indeterminate sex is beckoning from behind the gauzy curtains of a palanquin. A few dour fellows, some in loincloths and others in habits, pretend stoically not to notice it, while waiting to be noticed themselves. The whole lot look like they've been arguing among themselves right up until you opened the door.

In this situation, it is highly advisable to get an advance notion of whether a recipe for self-conquest holds more promise than acting tough while looking in the mirror and whether it is safer than a straightforward lobotomy. To that end, you should examine the meme that peddles it, and its senders, for some sign of understanding of what a self is and how it works. As with any unfamiliar meme, it is also advisable to question its motives, lest a modest bid for self-renovation ends up with the cognitive equivalent of your home being overrun by the Huns. Having gotten this far in my book qualifies you for both those tasks. As promised in the preface, I have lined up for you the conceptual tools needed to understand how the mind really works. Armed with these tools, you can now start figuring out what's best for you, and the self-knowledge you stand to gain can also help you fend off predatory memes.

Self-knowledge has been highly valued by many philosophical and religious traditions, East and West. Laozi's comment, "Knowing others is wisdom; knowing the self is enlightenment," and the Buddha's gospel of self-knowledge would not have sounded out of place in Greece, where the entrance to the shrine of Apollo's Oracle at Delphi bore the injunction: "Know thyself."[7] Such wide endorsement bodes well for the idea of self-knowledge in the abstract, but it is in the details where, as the saying goes, the devil resides. Insofar as they are rooted in intuition or revelation, the actual bits of knowledge that such traditions offer by way of following up on their own advice are like sausages from a street-corner cart: they contain meat, but also other ingredients, some of which resist ready identification—even if (or perhaps especially if) the vendor claims to answer to a higher authority.[8]

It took humanity a long time to come up with a trustworthy procedure for monitoring the origin and quality of the ingredients in the sausage of knowledge. One of the clearer signs that such a procedure was forthcoming was the establishment in November 1660 in London of the world's first academy of sciences. The twelve founders who met at Bishopsgate decided to form a "Colledge for the Promoting of Physico-Mathematicall Experimentall Learning," which was incorporated two years later as the Royal Society of London. The motto of the Royal Society, *Nullius in Verba* (taken from a line by Horace, which translates roughly as "no faith in mere words"), affirms its commitment to the pursuit of objective knowledge through a process that used to be called "experimental philosophy" and is now called simply "science."[9]

The scientific process ensures that quality control over knowledge is reliable and transparent by placing it in the hands of a community of people who engage in theoretical inquiry, empirical studies, mathematical formulation, and open debate. Science is participatory: anyone who can demonstrate some kitchen skills is free to bring their own stuffing (mystery meats excepted) and learn sausage making, or try to convert sausage eaters to masa tamales or rice-noodle spring rolls. By taking note of the origins of the ingredients, sharing the recipe, and making the finished product public, science makes for the safest knowledge one could wish for.

Over the course of the past several decades, cognitive science has served up a full menu of quality self-knowledge, beginning with the computational fundamentals of representational processes, through theories of perception and action, memory, thinking, and language, to insights into personality, social cognition, and consciousness that arises from the dynamics of effective, narrative, and phenomenal Selves. The understanding of the human mind that has been gained, of which this book has offered only a glimpse, has a direct bearing on what emerged earlier in this chapter as the key happiness-related issue: the apparent need to choose between world conquest and self-conquest. As we shall presently see, the new understanding exposes both these choices as ultimately self-defeating.

Peace Is Struck in the Republic of Soul

Let us take another look at the world conquest route first. The example I chose to illustrate it with earlier, that of Alexander the Great, may have seemed far-fetched for most of us. A middle-class hero (let alone working-class hero) is something to be, but what could it possibly have to do with a dead Greek king? In my defense, I'll quote again from Whitman's *Song of the Open Road*: "Now understand me well—It is provided in the essence of things, that from any fruition of success, no matter what, shall come forth something to make a greater struggle necessary."[10] All of us are susceptible to the same bug that drove Alexander up the wall of Asia; only the severity of the affliction differs. In claiming this, I am not appealing to Whitman's authority—it is just that his insight happens to coincide with what cognitive science tells us about the dynamics of motivation.

The "hedonic treadmill" on which we are plonked at birth ensures that the perceived returns from equal increments of achievement will continually diminish. At the same time, a course of action grows increasingly difficult to set aside, as long as it meets with consistent success, no matter how dwindling; past reward history pushes people to persevere.[11] The combination of these basic traits of human nature sets up the dedicated world-conqueror types for a rat race that they cannot abandon and cannot ultimately win. The tragically minded among them would find a certain consolation in Rilke, who wrote that "the purpose of life is to be defeated by greater and greater things."

I seriously doubt that life has any purpose at all, but that's not an excuse to waste it on striving for more and more of the same. The "greater struggle" in Whitman's adage becomes your lot only if you want more and more of what you already have. Strike world conquest, then. This seems to leave us with the other option, that of self-conquest, which is not a cakewalk either, by any account.[12] The particularly tricky part about self-conquest is how to bring it about with as little destruction as possible. (The idea, I guess, would be to become like Diogenes, but with enough sense left not to urinate on people you don't like.) There is a price for every victory in any war, and a civil war is no exception. Even if you don't mind carpet-bombing your neighbors before venturing out to conquer them, doing the same to yourself would really bring home the meaning of "collateral damage." Then there is the old Soviet-era joke in which the legendary Radio Yerevan (no relation to the actual Armenian broadcasting service) is asked, "Will there be a World War III?" and replies, "In principle, no. But the struggle for peace will be so great that not a house will be left standing."

So, is there change we can believe in? The answer, I suggest, is yes, and not just in principle. We did learn a thing or two about selves over the past couple of millennia. As we now know, a regular human self is part virtual construct, part distributed entity, with the latter best thought of as a network of cause and effect that transcends the boundary between the individual and the environment, which includes the society and the material world.

This crucial piece of self-knowledge casts the old notions of world conquest and self-conquest in an entirely new light.

It is a mistake to think of self-conquest as the subjugation of a mindless horde of wild emotions and desires by a disciplined, perfectly reasonable tin-soldier homunculus that holds an exclusive title to mindhood. What your mind is really made of is perceptions colored by emotions (and emotions evoked by perceptions), actions initiated by desires (leading in turn to other desires), reasoning influenced by beliefs (giving rise to new beliefs or driving away old ones), and decisions pushed around by the computational avatars of your family, friends, teachers, and idols (and pushing back at them). None of the actors in this game are persons (just as a car wheel is not a vehicle); they are all more or less special-purpose computational processes that can only attain full human mindhood collectively as the whole shebang goes about its business.

The mind's being distributed and collective places certain constraints on what a self can and cannot do in striving for change and hints at the likely consequences of certain courses of action. Build a wall to keep your barbarians out, and you end up besieged, excluded, divided, and diminished at the same time. Keep them in and teach them to pay taxes and vote, and everybody wins—that "everybody," of course, being you. (I refer here to the indigenous barbarians, as opposed, say, to the Huns, who really ought to be kept out. Even some folks who you'd think are civilized should be treated with suspicion: remember Greeks bearing gifts.)

Because the mind of an individual is computationally entangled with those of others in his or her social circle, minds are distributed not only across many processes running inside their respective "home" brains but also over a sizable chunk of the world that surrounds them. Thus, the distributed, collective nature of the mind constrains not only self-conquest but also world conquest. The struggle to bend the world, or even just a small corner of it, to one's own will necessarily remakes the would-be conqueror's self.

One often encounters sentiments analogous to this one in geopolitical thinking. Empire-builders realize sooner or later that their home country is being transformed by its overseas dominions; occupation is known to corrupt the occupier even as it grinds down the occupied. My extension to individuals of such observations, which normally apply to nation-states, capitalizes on the modern understanding of the distributed, collective self by cognitive science, but it also echoes David Hume's brilliant insight comparing "the soul . . . to a republic or commonwealth," as recounted in Chapter 3.

Much in the same way as the relationship between minds and computation captured by the "computer metaphor" proved to be literally true, Hume's vision of the mind as a commonwealth has been transformed by cognitive science from a mere metaphor into an empirically substantiated computational theory. Hume's outstanding explanatory move, no longer metaphorical, still offers a valuable and vivid picture of how a mind (or "soul") can act cohesively despite being distributed, and how a distributed being can undergo change while its identity endures.

Less astute philosophers than Hume tended to picture the mind as a compartmentalized black box, whose emergent personality reflects the balance of terror in the strife between the compartments. An unfortunate consequence of this view has been the notion—still endemic in some schools of personality and clinical psychology—that the pursuit of happiness must involve a violent realignment, or, as I put it earlier, conquest. On the one hand, the happiness-craving black box is supposed to wage war against itself—a campaign that is bound to meet with resistance.

(If you ever tried it, you'll know why.) On the other hand, war must be waged against others to induce them to further what for them is someone else's happiness agenda; this too is unlikely to contribute to peace in the neighborhood.

The black-box self in this mistaken view of how the mind works finds itself in a perpetual state of war of its own making, not unlike the assorted cold warriors in Stanley Kubrick's film *Dr. Strangelove, or How I Learned to Stop Worrying and Love the Bomb*. I am thinking here in particular of the personage of Dr. Strangelove himself, who has to fight off his own hand that tries to strangle him when he is not scheming about how best to fight the Russkies (who, of course, fully reciprocate).

Cognitive science's discovery that minds are in fact not sealed black boxes but open, intermingled societies of computational processes gives one hope that a person wishing for a happier life can attain it through gradual, cognitively transparent change.[13] A unique feature of an open society—Hume's "republic or commonwealth"—is its ability to change itself, and sometimes its neighbors and partners, to the better by nonviolent means. To my mind, Hume's ideas and the lessons from cognitive science that corroborate them suggest that it should be possible for people to strive for happiness without resorting to any kind of "conquest." It is interesting to note that whereas the classical recipes for collective happiness predicate the betterment of a society on the self-improvement of its members, what we just discovered is that these two kinds of processes may be expected to follow similar dynamics.

This dynamics is anything but simple: people's minds, and even more so societies made of people, are very complex systems. The path of individual or societal change is thus rarely short, straight, or smooth. An open society may nearly succeed at suicide, only to reemerge, after unspeakable violence to self and others, as one of the most progressive forces on the planet. (Consider Germany, from the Nazi win in the 1933 elections to the 1967 shooting of a peaceful demonstrator by the police and the ensuing public outcry, which eventually transformed the still-conservative postwar country.[14]) Or it may resort to internal violence to initiate the emancipation of its downtrodden second-class citizens, yet complete this emancipation by peaceful means. (Consider the United States, from the judicial murder of John Brown in 1859 to the election of President Barack Obama in 2008.)[15]

It looks, therefore, like there are reasons for optimism. Martin Luther King Jr. once said that "the arc of the moral universe is long but it bends toward justice." But cognitive science tells us that the moral and cognitive dimensions of the human mind are coextensive. Similarly, the idea that the moral progress of a society is driven by the cognitive betterment of its members—which, unsurprisingly, brings about also increased individual well-being—is implicit in Aristotle's concept of practical wisdom, or "phronesis," and in the "right mindfulness" that is part of the Buddha's Eightfold Path.[16] Thus, with any luck, the lifelong psychological war of conquest that each of us has been waging in the name of personal happiness and prosperity will end with a velvet revolution.

Ulysses Leaves Ithaca Again

Cognitively transparent (hence peaceful), gradual self-change of the kind that promotes well-being and, indeed, happiness is helped along by the accumulation of experience. That life experience is good for your practical wisdom has been noted by philosophers; more importantly, this notion turns out to be very much along the lines of what science has learned about the role of experience in cognition.[17]

The idea that experience is central to cognition has been with us since Chapter 2, where I sketched a rational foundation for using the past to anticipate the future: the amazing Bayes Theorem. In subsequent chapters, we saw that the buildup of experience enriches and refines all cognitive functions: perception, motor planning, problem solving, decision making, and language. It is a truism that the sum total of a person's episodic memories constitutes a large part of his or her persona—what I called the narrative Self.[18] We should remember that the other part of the mind's *I*—the phenomenal Self—is swept along by the same current: the present is always experienced through the memories of the past.

There are good reasons why the accumulation of experience feels good and why it promotes happiness (in a reasonably well-adjusted self, that is; as it usually happens, it is the rich who get richer—but you did not need me to tell you that, did you?).[19] The urge to explore, accrue information about the world, and use it to dodge the "slings and arrows of outrageous fortune" (or catch such stuff as looks good) makes evolutionary sense. Feeling good about mastering novelty through learning appears to be the currency with which the brain is bribed into leaving the couch and venturing outside. Even though what evolution *acts* on is not fleeting good moods but lasting good outcomes, in its vocabulary "lasting" means "lasting long enough to procreate profusely"—what a lovely way to burn! Thus, feeling good is the means, but the end is happiness. Could you imagine it any better than that?

To sum up, we now know enough about how the mind works to be able to explain not just *why* happiness can be increased by the pursuit of experience, but *how* it happens. In a drastically condensed form, the complete explanation reads as follows. The world is partially predictable. Predicting the future requires remembering the past. Cognitive systems, of which the human mind is a prime example, use accrued past experience to their advantage by resorting to statistical inference. Such inference ranges from simple extrapolation of a few variables by regression on the available data to the construction of complex virtual-reality models intended to simulate the behavior of objects and the unfolding of events, including other minds and social interactions. For their pursuit of experience in the service of these computational needs, systems that are subject to evolutionary pressure, as the human mind is, are rewarded, both in real time and in the long run. This, then, is the entire account of the happiness of pursuit.

How does my quest for an algorithmic understanding of happiness measure up against the explanatory excitement fanned by all the talk about computation in this book? I shall not pretend that the understanding at which we have arrived spells out a comprehensive algorithm for leading a happy life (although it does suggest some well-motivated actionable ideas, which could easily fill another book). No, its main value lies elsewhere.

Imagine that your search for spiritual solace brings you to the mountain retreat of a widely revered sage. The hermit, who looks reassuringly radiant and serene, gladly reveals the secret: ubik. You ask for details and are told that ubik is something that all people have (or perhaps merely think they have—the account, which mainly takes the form of parables, is unclear on that point), and that it is generally a good idea to begin with getting to know your ubik, which in turn will enable ubik-improvement, with all the ensuing benefits. You are too polite to press the issue further and so take your leave, hoping to work it out on your own.[20]

This vignette is, of course, prompted by the classical recipe that predicates happy life on self-knowledge, self-improvement, and, eventually, selfless conduct, yet stops short of explaining what kind of thing it is, exactly, that you need to come to know, improve, and perhaps transcend. This used to be a singularly frustrating situation, but times have changed. Instead of merely exhorting you to seek self-knowledge, this book explains what a self is, so that you may know yours better. One classical motif does, however, persist: writing the sequel for your story is still up to you.

I conclude mine by revisiting Tennyson's poem *Ulysses*, from which I quoted back in Chapter 3. The hero—back in Ithaca after ten years on the fields of Troy and another ten trying to get home, through many adventures and much adversity—is called by an irresistible call to depart. It takes Tennyson's Ulysses seventy lines, some of the very best in English verse, to explain why. Of these, I always found the last one the most striking, and it is that line that I will now tweak to make my point (with impunity, as all persons concerned have long since sailed beyond the sunset). My version of the closing line puts a little less value on resolve and tenacity, which all heroes have in abundance, and a little more on the twin virtues of recognizing when a cycle of experience has come to a completion, and knowing what to do then:

To strive, to seek, to find, and ~~not~~ then to yield.

Although this may sound like the end of the present story, we should take it for what it really is: a new beginning. The happiness of pursuit is such that there is always more of it to be had, as long as we remember that the moment of attainment is the perfect time to start striving for something new—but not until after we pause to savor the moment, which is what I am going to do now.

Ithaca, NY
March 2011

ALWAYS COMING HOME[†]

We shall not cease from exploration
And the end of all our exploring
Will be to arrive where we started
And know the place for the first time.

— T. S. ELIOT, "Little Gidding," V, *Four Quartets* (1942)

Ever since Jorge Luis Borges and Stanisław Lem perfected the art of reviewing nonexistent books, thereby setting a new standard for insouciant sophistication, aspiring authors who are at least as ambitious as they are lazy have been faced with a choice: to actually write the book of their dreams, or to pretend that it has already been written and proceed directly to write a review for it. An enlightened person—one who has cultivated a due appreciation for the self-limiting nature of any action (a quality that is virtually unknown among aspiring authors)—would opt for neither

---

[†]The title of this review of *The Happiness of Pursuit*, which is conveniently included with the text itself, has been borrowed by its author from Ursula K. Le Guin (1985).

of these two alternatives. I, who am the nominal author of *The Happiness of Pursuit* (*THoP*, pronounced "tea-hop"; Edelman 2012) as well as of the present review, have chosen *both*.

Although such lapse of judgment is not without precedent, an excuse for it is still needed. The precedent I have in mind is Lem's volume *A Perfect Vacuum* (1971/1999), whose opening chapter is a mind-numbingly virtuoso review of the book of which it is part (and which otherwise reviews works that are strictly nonexistent), as well as a trove of excuses for its own existence. Mine derives from the observation that the earlier version of me, which had set out to write *THoP* a couple of years ago, and the later version, which has finished it and is now well into the second paragraph of this review, are not one and the same person. Time and experience change people, and a protracted struggle with a book is bound to take its toll on the writer, even if in the end it is the writer who wins. While the present reviewer and the author of *THoP* may be related to each other (in the same sense that Ulysses who is being ferried back to Ithaca by the Phaiakians is related to the younger Ulysses at Troy), they cannot be identical. As Heraclitus would have agreed, the same you cannot step twice into the river.

The author of *THoP* seems to have understood that. Although in the end he sides with Tennyson's hero, who is about to set sail again, his version of Tennyson's key line does not rule out the possibility of another happy return to Ithaca by a latter-day Ulysses, yet again changed by his voyages. One is left to wish that *THoP* were a bit more explicit about that.

In the psychology of mind, the idea of a changing, growing self that is being constantly shaped by experience has its roots in the radical empiricism of Ernst Mach (1886) and William James (1912). Given how central this idea turns out to be in *THoP*, it is a pity that the author did not do more to place it in an appropriately developed historical context. Doing so might have imparted a more meaningful narrative structure to this loosely connected series of essays, which could have in turn supported an inclusive survey of the relevant scientific literature instead of the haphazard selection that did make it into the book's endnotes. Even more importantly, an orderly exposition of the science behind the book's arguments would have probably prevented crucially important points from being relegated to the endnotes, as when the discussion of multiple realizability of phenomenal experience—a question on which both consciousness and happiness, human and machine, are predicated—got compressed into one sentence, buried in an end-of-chapter remark (Chapter 6, note 26).

Despite the promise we were made back in the author's note all those pages ago, to "[set] aside the conventional divisions between science and the humanities," the book's literary and other cultural examples are as scattershot and idiosyncratic as the scientific references it offers. Instead of requiring the bewildered reader to mentally juggle Homer, Shakespeare, flying marmots, cameo appearances of movie comedians, lovesick yeast, and whatnot, the author should have staked out a well-defined, unitary cultural reference space to augment the space of scientific concepts that he set out to map. Naturally, one is prompted to conclude that the idea of writing up something along the lines of *The Happiness of the Simpsons* had failed to indulge the author's pretensions to erudition and his addiction to the absurd, the signs of which crop up all too often in his writings.

Conspicuous in its omission from *THoP* is a philosophical take on the issues of interest—an omission that is all the more inexplicable in the light of the key role played by philosophy, throughout the ages, in humanity's attempts to make sense of its own existence. Granted, the author has announced elsewhere (Edelman 2008c, p. 262) his allegiance to Quine's idea that there should not be first philosophy, yet he seems to be not entirely consistent in this stance, quoting from philosophers he approves of when it is convenient and glossing over those of whom he does not. His choice of philosophers is, incidentally, skewed toward the East (Hume and Wittgenstein both being increasingly seen nowadays as espousing Buddhist views; cf. Kalansuriya 1993; Scharfstein 1998)—a curious inclination, seeing that in the cultural domain the author's examples are all Western. (For instance, out of the eight literary works listed in Chapter 1, note 4, seven originate in European or American culture.) I wonder whether or not there is still something that he can do to avoid being charged with Orientalism.

On reading over these pages, I fear I have not called sufficient attention to the book's many virtues. It includes a number of fine distinctions. (A few of those are clearly borrowed from classical sources, such as the penultimate paragraph on p. 180, which is from the conclusion of Borges's (1935/1962a) review of *The Approach to Al-Mu'tasim* by Mir Bahadur Ali, but then Borges himself, in that very review, refers to such borrowing practice as "honorable.") I shall refrain from expounding on those virtues, lest I be accused of undue partiality toward the author of *THoP*.

Indeed, seeing that he and I are related, I feel I have to make an extra effort to prove that I am not doing his bidding. To that end, I shall reveal a little secret that I inherited from him: when he set out to write *THoP*, he had been motivated solely by what he perceived as a compulsion to create a book-sized home for the mangled quote from Tennyson's *Ulysses* with which *THoP* ends—a line that he had thought would in itself be a worthy, yet too elliptical, message to the reader.

The experience of writing *THoP* has changed his—my—mind: the book's message, along with its real ending, is, I now believe, wholly contained in Christopher Logue's poem with which it opened:

You ask me:
What is the greatest happiness on earth?
Two things:
changing my mind
as I change a penny for a shilling;
and
listening to the sound
of a young girl
singing down the road
after she has asked me the way.

ACKNOWLEDGMENTS

The quartet of questions reproduced in the author's note were jotted down while I was hiking down Harris Wash, a tributary of the Escalante River in Utah. For paper, I used the back of a receipt from Escalante Outfitters, the best outlet for topo maps, local beer, and homemade sausage and goat-cheese pizza in Escalante, which is the Utah equivalent of Panamint Springs, California. There is no place like the great deserts and canyonlands of the American Southwest for keeping one's mind focused on the really important things in life.

Returning to the alternate reality of this book—Julie Simmons-Lynch deserves special thanks for reading and commenting on early drafts of several of its chapters. I realize that being an early adopter has its disadvantages, and I appreciate her daring.

I thank all my friends from Ithaca for making it such a great place to play with ideas—even if some of those do not pan out. (Here I have in mind the intriguing, yet ultimately barren, idea, which came up in Chapter 2, suggesting that the ferocity of a laboratory animal might have something to do with its performance as a ballistic missile. I borrowed it from Dr. Larry Taylor, via my esteemed Cornell colleague Barb Finlay.) A discussion with another colleague, Alice Isen, helped me think about the relationships between happiness and cognitive processing (Chapter 7). Not all such discussions happen on campus: James Cutting told me about the link between dopamine receptor properties and long-range human migration patterns (Chapter 7 again) while we were riding a chair lift at our local ski hill. I find skiing with people in the hope that some of their knowledge will rub off on me to be a fun way of combating my scientific narrow-mindedness, and Ithaca happens to be a good place to indulge in this mode of self-improvement.

Given my predilection for extracurricular activities, I would probably never have started working on this book if it were not for Ben Mauk, who is responsible for what proved to be an effective initial impetus for it, and Björn Merker, who generously shared with me his thoughts on world conquest and self-conquest, thereby helping me convince myself that the last paragraphs of Chapter 7 (which were the first that got written) were worth expanding into a book. I thank Jim Levine for finding a good home for my manuscript, and I am also very grateful to my editor at Basic Books, T. J. Kelleher, for timely encouragement, which helped me bring this work to completion, and for good advice, which made completing it so much more enjoyable.

Finally—the pursuit of this project (and of much else) would have been devoid of happiness were it not for Esti, Ira, and Itamar, to whom I owe more than I can ever thank them for.

NOTES

NOTES TO CHAPTER 1

1. By 2003 or so, it became apparent to me that the most recent available conceptually unified and convincing PSYCHOLOGY TEXTBOOK was *Human Information Processing* by Lindsay and Norman (1972). I now teach cognitive psychology from a book titled *Computing the Mind: How the Mind Really Works* (Edelman 2008a).

2. I refrain from defining HAPPINESS and refer the reader instead to a characteristically informative and entertaining overview of what philosophers, scientists, and normal people usually mean by it, offered by Daniel Gilbert in his book *Stumbling on Happiness* (2006), pp. 31–38. Gilbert remarks that "asking what happiness *really* is . . . is approximately equivalent to beginning a pilgrimage by marching directly into the first available tar pit."

3. Some of the literary examples that come to mind in connection with the idea of LIFE AS A JOURNEY AND A HOMECOMING are: *The Odyssey* by Homer; *Journey to the West* by Wu Chengen; the *Josephus* trilogy by Lion Feuchtwanger; *Magister Ludi* by Hermann Hesse; *The Hobbit, or, There and Back Again* and *The Lord of the Rings* by J. R. R. Tolkien; and the *Wizard of Earthsea* cycle and the unique gem of a book titled *Always Coming Home* by Ursula K. Le Guin.

4. John Keats used the phrase "SOUL-MAKING" in his journal-letter to George and Georgiana Keats, April 21, 1819, reprinted in Strachan (2003). The idea that human flourishing, or eudaimonia, is intimately

connected to human nature has been discussed at length by Aristotle in *Nicomachean Ethics* (350 B.C.E.). For a modern perspective, see Irwin (1991) and Nagel (1972).

5. Many of the DETAILS that I have omitted can also be found in my earlier book, *Computing the Mind* (Edelman 2008a).

NOTES TO CHAPTER 2

1. Among the books that make a point of revealing the most important "secret" of COGNITIVE SCIENCE to their readers are: Warren McCulloch's *Embodiments of Mind* (1965), the definitive collection of McCulloch's academic papers and essays, including the famous 1943 paper with Walter Pitts that establishes an equivalence between Turing machines and networks of formal neurons; Marvin Minsky's *The Society of Mind* (1985), a profoundly insightful speculation on the nature of the mind by one of the pioneers of artificial intelligence; Drew McDermott's *Mind and Mechanism* (2001), a modern look at natural and artificial intelligence; and my own *Computing the Mind* (Edelman 2008a).

2. A computational analysis of the LIGHTNESS PERCEPTION problem can be found in Adelson and Pentland (1996).

3. The story of THE OMEN AT AULIS is told in book 2 of *The Iliad*, lines 301–332. The main events described in *The Iliad* unfold over about seven weeks of real time, but most of it deals with just four days.

4. An account of ARISTOTLE being confounded by the story of the interpretation of the omen at Aulis is given by Marcus Tullius Cicero in Wardle (2006), p. 283.

5. CAUSATION is in principle problematic but in practice well founded. A definitive early treatment of causation—both its philosophical problematicity and its empirical resolution through statistical inference—was offered by David Hume in *A Treatise of Human Nature* (1740). The full text of the *Treatise* is available online courtesy of Project Gutenberg.

6. Because "nothing in biology makes sense except in the light of EVOLUTION" (Dobzhansky 1973), this book of necessity resorts to evolu-

tionary arguments. In appealing to evolution as an explanatory framework, I adopt Eva Jablonka and Marion Lamb's (2007) extension of the standard Darwinian "modern synthesis":

> In *Evolution in Four Dimensions* (2005) we identify four types of inheritance (genetic, epigenetic, behavioral, and symbol-based), each of which can provide variations on which natural selection will act. Some of these variations arise in response to developmental conditions, so there are Lamarckian aspects to evolution. We argue that a better insight into evolutionary processes will result from recognizing that transmitted variations that are not based on DNA differences have played a role. This is particularly true for understanding the evolution of human behavior, where all four dimensions of heredity have been important.

7. The discussion regarding "WHOSOEVER IS WISE?" is found in the Bavli Talmud (Mishna, Seder Kodashim, Masekhet Tamid 32:1).

8. The head-banging habits of MOLE RATS have been reported by Kimchi, Reshef, and Terkel (2005).

9. An informed, insightful, and engaging treatment of the EVOLVABILITY OF FORESIGHT is offered by Daniel Dennett (2003a) in his book *Freedom Evolves*. For a review of the relationship between brains, innovation, and evolution in birds and primates, see, for instance, Lefebvre, Reader, and Sol (2004) and Lefebvre and Sol (2008).

10. The BAYESIAN approach to understanding the mind is the subject of numerous books (for example, Glymour 2001; Knill and Richards 1996) and scholarly articles (see Chater, Tenenbaum, and Yuille 2006; Tenenbaum and Griffiths 2001). Colin Howson and Peter Urbach (1991) explain why Bayesian reasoning is indispensable in science.

11. Let me stress again, as I did earlier, that to replicate precisely the mind generated by a particular brain one must reproduce the TRAJECTORY DYNAMICS of that brain's states as they unfold over time (Chalmers 1994), including, crucially, the intrinsic categorical structure of the space of possible trajectories (Fekete and Edelman 2011). As the great American philosopher Charles Sanders Peirce (1868, p. 149) noted, "At no one instant in my state of mind is there cognition or representation, but in the relation of my states of mind at different instants there is."

NOTES TO CHAPTER 3

1. Francis Crick's (1994) quip, "You are nothing but a pack of neurons," which is a paraphrase of an exclamation from *Alice's Adventures in Wonderland*, has been updated recently to: "You are nothing but a pack of COMPUTATIONS" (Edelman 2008a, p. 500).

2. According to Philip K. Dick's theory of LIFE ORIGINS, "eons ago, in the remote past, a bit of inanimate matter had become so irritated by something that it crawled away, moved by indignation" (Dick 1954/1993).

3. In processing PHEROMONE GRADIENTS, yeast cells perform close to the absolute physical limit of detection (Endres and Wingreen 2008).

4. The FIGHT SCENE is from Shakespeare's *Romeo and Juliet*, act III, scene I.

5. The no. 1 tool, which reveals computation that is in the nature of things, makes you a bit like Neo in *The Matrix*, who can perceive the flowing numbers that make up reality. The no. 2 tool is akin to what Daniel Dennett (1987) calls the intentional stance in cognitive science. The no. 3 tool is related to Herbert Simon's (1973) formulation of hierarchical abstraction as an explanatory move and to the multiple levels of analysis of information-processing systems introduced by David Marr and Tomaso Poggio (1977).

6. Some general DESIGN CONSIDERATIONS for cognitive systems can be found in Sloman (1989) and Sloman, Chrisley, and Scheutz (2005).

7. For an intriguing treatment of SUPER POWERS, see Saunders (2008).

8. Overcoming clinical depression, whose symptoms can include a persistent and general LACK OF MOTIVATION, may require professional help.

9. I follow Minsky (2006) in holding that "EMOTIONS are different Ways to Think" (cf. Sloman et al. 2005).

10. Our first encounter with Hume's *A Treatise of Human Nature* (1740) was in the note on CAUSATION in Chapter 2 (note 5). The passage comparing the soul to a republic is in book I, part IV, section VI, paragraph 19.

11. My own favorite example of the importance of INFORMATION PROCESSING IN A WAR EFFORT, from the days before military intelligence was

branded as a contradiction in terms, is the Bletchley Park operation. Bletchley Park is a country estate in Buckinghamshire, England, which during World War II housed the cryptanalysis section of the British military intelligence. If it were not for the daily successes of the code-breaking effort carried out at Bletchley Park by an illustrious team that included Alan Turing and I. J. Good, you would not be reading this note. The historical-causal chain in this case is exceptionally clear. Had the radio communications between the German submarine fleet in the Atlantic and its headquarters not been deciphered, the Allied fleet would have been unable to hunt down the submarines; the shipments of vital war supplies from the United States to Britain would have been disrupted; the Battle of Britain, which for many months hung in the balance, would have been won by Germany; the newly freed resources would have given the German army a critical boost on the eastern front; and my Russian-Jewish parents would not have survived the war, let alone met, fallen in love, and had a son with a predilection for historical digressions. Yet the present digression is not entirely gratuitous; for a connection between Bletchley Park and computational neuroscience, see Gold and Shadlen (2002).

12. Note that Jeremiah's observation on FOOLISH PEOPLE (5:21) applies recursively to constituents of minds.

13. The discovery of STRUCTURE IN INFORMATION gathered by its senses is a developing mind's first step toward learning the world (O'Regan and Noë 2001; Philipona et al. 2004).

14. One wonders whether Romeo might not have been better off in the long run by KILLING BENVOLIO instead of Tybalt.

15. The literature on FACE RECOGNITION is vast. Here are a few useful entry points into its main subdivisions: psychophysical (Diamond and Carey 1986; Moses, Ullman, and Edelman 1996; O'Toole, Edelman, and Bülthoff 1998), neural (Eifuku et al. 2004; Haxby et al. 2001; Rolls et al. 1989; Young and Yamane 1992), and computational (Edelman 1998; Edelman, Reisfeld, and Yeshurun 1992; Lando and Edelman 1995; O'Toole and Edelman 1996).

16. Individual neurons in sheep's brains that are tuned to SHEEP FACES have been reported by Kendrick et al. (1996) and Kendrick and Baldwin (1987).

17. For an integrated treatment of the problem of OBJECT RECOGNITION, see Edelman (1999).

18. This VIRTUAL LINEUP method was implemented by Jungman, Levi, Aperman, and Edelman (1994).

19. In general, GRADED-SIMILARITY COMPUTATION works very well with neurons. The array of numbers that form a snapshot of a face (or of any other object) in a biological visual system is, very straightforwardly, a pattern of activity of some neurons (specifically, the activities of those neurons that respond to a face *are* the coordinates of the face-space point that represents it). This array of numbers can be represented collectively, in a very compact form, by another neuron, to which their outputs are fed. For each stimulus category, the brain only needs to store a few snapshots or exemplars; others can then be estimated by interpolation (Edelman 1999), which can be implemented as Bayesian regression (Bishop 2006)—the normative approach to learning from experience.

20. The actual lines are:

But, soft! what light through yonder window breaks?

It is the east, and Juliet is the sun. (act II, scene II, lines 2–3)

21. A discussion of ANALOGY AS THE CORE OF COGNITION can be found in Hofstadter (2001). For an application to vision, see Edelman and Duvdevani-Bar (1997).

22. For a computational analysis of the predicament of a mind that does not trust REALITY, see Edelman (2011b).

23. In vertebrates, a functional equivalent of the GRAND MAP is found in the superior colliculus, a midbrain structure (Merker 2007).

24. The concept of AFFORDANCES was introduced by J. J. Gibson (1979); discussed recently by Alva Noë (2004, p. 105), who wrote: "To perceive . . . is to perceive structure in sensorimotor contingencies"; and reviewed in Edelman (2006). Regarding basilisk lizards, see Glasheen and McMahon (1996).

25. Konrad Körding and Daniel Wolpert (2006) describe DYNAMIC SIMULATION in the context of Bayesian motor control.

26. Computational aspects of BEAUTY have been discussed in Edelman (2008a, ch. 5) and Schmidhuber (2008).

27. Empirical studies of SCENIC BEAUTY are rare (Daniel 1990). Recent work suggests that physical factors (such as complexity, openness, and water features) do a better job of accounting for judgments of scenic beauty than biome category. Still, Ke-Tsung Han (2007, p. 551) concludes: "It appears that no single current theory alone can sufficiently explain the causal processes responsible for any consistently favorable reaction to natural settings in general and to biomes in particular."

28. Tolkien (1954), p. 54.

29. Morris (1971), p. 138.

30. As Robert A. Wilson (1981) expresses it in *Masks of the Illuminati*, "To the puer, all things are puella." The influence of BODILY STATES such as thirst on perception has been documented, for example, by Melissa Ferguson and John Bargh (2004).

31. For an overview of NEUROECONOMICS, see Glimcher et al. (2008).

32. Shakespeare, *Romeo and Juliet*, act I, scene II.

33. The quote is from Tennyson's poem *ULYSSES* (1842), line 51.

34. For a natural history of PARADISE, see Manuel and Manuel (1972). Björn Merker's "Vehicles of Hope" (1993) offers a book-length treatment of related issues. The preponderance of belief in "heaven" (shared by three-quarters of the population of the United States) is reported by Greeley and Hout (1999). Interestingly, there seems to be no correlation between this belief and education or socioeconomic status (Flannelly et al. 2006).

35. For a science-fictional treatment of the possibility for a person to reach in and adjust his or her mind for PERMANENT CONTENTMENT with an ongoing activity, see Greg Egan's *Permutation City* (1994). The resulting creature may be happy, but it is no longer human.

36. Gay Watson's "Buddhism Meets Western Science" (in *Religion and the Brain* 19, 2001) is an extremely brief introduction to the PSYCHOLOGY OF BUDDHISM. On a humanist reading, the Buddhist philosophy of mind, ethics, and enlightenment (Gier 2002, 2007; Gier and Kjellberg 2004; Siderits 2007) suggests that "awakening does not free one *from* the world; it frees one *for* the world" (Garfield and Priest 2009, p. 76).

37. The original quote reads: *Цель—это только средство. . . . Счастье не в самом счастье, но в беге к счастью.*

NOTES TO CHAPTER 4

1. The only one described in *The Iliad* as "HAPPY," and then too in the past tense, is Priam, the king of Troy and the father of Hector. While allowing Priam to take the body of Hector back to Troy for a burial, Achilles says, "And you too, O Priam, I have heard that you were aforetime happy. They say that in wealth and plenitude of offspring you surpassed all that is in Lesbos, the realm of Makar to the northward, Phrygia that is more inland, and those that dwell upon the great Hellespont; but from the day when the dwellers in heaven sent this evil upon you, war and slaughter have been about your city continually" (Book XXIV).

2. For a discussion of the "cosmological and ideological aspects" of the SHIELD OF ACHILLES, see Philip Hardie's paper, "Imago Mundi" (1985). If you are visiting from another planet and have not yet made up your mind what to think of this one, don't read W. H. Auden's poem "The Shield of Achilles" just yet.

3. Ovid, *Metamorphoses,* Book XIII, 299–336.

4. As noted in *The Principles of Psychology* (James 1890, p. 608), William James borrowed the expression "SPECIOUS PRESENT" from E. R. Clay.

5. The MIRROR OF GALADRIEL scene is from *The Lord of the Rings*: *The Fellowship of the Ring* (Tolkien 1954, p. 469). One might guess that it reflects Tolkien's sensibilities as a Catholic; the issue of free will (which is more complex than people tend to think; for a modern treatment, see Wegner 2004 and Edelman 2008a, ch. 10) crops up in many other places in his work.

6. The ASSIMILATION of epigenetic information into the genome is explained in Jablonka and Lamb (2005).

7. Matteo Mameli and Patrick Bateson (2006) considered twenty-six different candidates for a scientific successor to the folk concept of INNATENESS and found none to be problem-free.

8. For some examples of the role of IMMEDIATE EARLY GENE expression in neural information processing, see Davis, Bozon, and Laroche (2003), Deisseroth et al. (2003), and Mayer, Watanabe, and Bischof (2010).

9. The CO-CONSTRUCTION OF THE ECOLOGICAL NICHE by its denizens is a central theme in Jablonka and Lamb (2005). For humans, important aspects of the niche are social; Drew Bailey and David Geary (2009) compare social competition to other factors in human evolution. The nature of the interaction of a species with its environment (including competition with its neighbors) is such that the ultimate challenge presented by the ecosphere to a species is open-ended—that is, it cannot be stated once and for all ahead of time (Clayton and Kauffman 2006).

10. The local synaptic LEARNING rule based on joint statistics of the activities of pre- and postsynaptic neurons was first envisaged by Donald Hebb in *The Organization of Behavior* (1949). Modern formulations focus on Spike Timing Dependent Plasticity (STDP) (Caporale and Dan 2008). Learning (of the kind afforded by such synaptic processes) supports the acquisition of novel behaviors in a variety of species, including gulls, who learn to drop clams onto hard surfaces to shatter them (Barash, Donovan, and Myrick 1975; Maron 1982), and New Caledonian crows, who can learn how to turn a piece of wire into a simple tool for extracting food out of a tight space (Hunt and Gray 2003). In humans, virtually all specific dispositions and behaviors are learned; among the few exceptions is the tendency to selectively associate snakes with fear (but not fear of snakes as such; DeLoache and LoBue 2009).

11. Paz, Gelbard-Sagiv, Mukamel, Harel, Malacha, and Fried (2010) recorded the activity of individual neurons in the HIPPOCAMPUS, an area of the brain implicated in navigation and in memory for locations and for sequences, while their human subjects viewed repeated runs of cinematic episodes. They found that within two to three presentations of an episode neuronal activity in a given time window became a faithful predictor of the activity to follow. The function of the hippocampus will be discussed in more detail later in this chapter.

12. The discovery that rats are capable of taking shortcuts through unfamiliar territory to get from one familiar place to another is due to Edward Tolman (1948), who hypothesized that this ability is supported by a kind of COGNITIVE MAP. Blind mole rats too turned out to be capable of shortcut-taking (Avni, Tzvaigrach, and Eilam 2008). The view of

the hippocampus as a cognitive map (O'Keefe and Nadel 1978) has been an important precursor of the modern theory that interprets its main function as that of EPISODIC AND SEQUENCE MEMORY (Iglói et al. 2010; Wood et al. 2000).

13. Concerning the role of location in word learning, Stephen Hockema and Linda Smith (2009, p. 471) write: "It seems that infants can build up strong associations between an object and its spatial location to the point where the spatial location can act as a surrogate for the object in labeling. (What is especially noteworthy about this is that the external binding through space seems to be developmentally essential for this progress to more complicated forms of binding to be made that no longer use space.) The point is this: Our environment has a consistent spatial structure, and our cognitive processes will make use of that regularity." More generally, the involvement of the HIPPOCAMPUS IN LANGUAGE PROCESSING is suggested by the finding that verbal shadowing (a test condition in which the subject is required to repeat aloud a stream of utterances heard through earphones) interferes with way-finding (Meilinger, Knauff, and Bülthoff 2008).

14. Mike Colombo and Nicola Broadbent (2000) have documented the involvement of the avian HIPPOCAMPUS IN SPATIAL TASKS. Both in food-caching birds (on chickadees, see Smulders, Sasson, and DeVoogd 1995) and in humans who are required to perform certain complex navigation tasks (such as driving a cab around London; see Maguire et al. 2000), the volume of hippocampal tissue correlates with the level of functioning. Interestingly, London bus drivers, who navigate the same labyrinthine city but adhere to fixed routes, do not exhibit the same correlation between hippocampal volume and employment duration as taxi drivers.

15. The involvement of the hippocampus in MENTAL TIME-TRAVEL into the (imagined) future is indicated by the inability of patients with hippocampal lesions to imagine new experience (Hassabis et al. 2007).

16. The study of EPISODIC AND PROSPECTIVE MEMORY in scrub jays described here was performed by Caroline Raby and her colleagues (2007). Uri Grodzinski and Nicola Clayton (2010, p. 983), who replicated those results, write: "The jays' remarkable ability to switch after

only one trial to caching the food they are currently sated on (i.e., against current motivation) suggests that caching is inherently directed towards needs that will be present at the future time of recovery."

17. James Russell, Dean Alexis, and Nicola Clayton (2010) report the gradual DEVELOPMENT OF PROSPECTION in young children.

18. The kind of training that can help reduce mind-wandering is MINDFULNESS MEDITATION (Baer 2003; Davidson 2010).

19. From Dante, *The Divine Comedy: Inferno*, canto V, line 121:
Nessun maggior dolore
Che ricordarsi del tempo felice
Nella miseria.

20. From the Robert Fitzgerald translation of *The Odyssey* (Homer 1998), book XV, lines 487–489. A similar sentiment is voiced by the long-suffering hero of Virgil's *Aeneid*, the Trojan exile who goes on to become the forefather of Rome. It is also shared by John Keats, who wrote: "Do you not see how necessary a World of Pains and troubles is to school an intelligence and make it a soul!" (from a letter to George and Georgiana Keats, February 14–May 3, 1819, reprinted in Strachan 2003).

21. For an empirical comparison of the HEDONIC value of material and experiential goods, see Nicolao, Irwin, and Goodman (2009) and Van Boven and Gilovich (2003).

22. C. P. Cavafy, *Collected Poems*. Translated by Edmund Keeley and Philip Sherrard. Edited by George Savidis. Revised Edition. Princeton University Press, 1992.

NOTES TO CHAPTER 5

1. The vole findings are by Young and Wang (2004). More generally, the core of a SOCIAL BEHAVIOR NETWORK comprises six areas in the basal forebrain and midbrain. This network has been implicated in the control of "aggression, appetitive and consummatory sexual behavior, various forms of communication, social recognition, affiliation, bonding, parental behavior, and responses to social stressors" (Goodson 2005, p. 12).

2. The word MEME originated in *The Selfish Gene* (Dawkins 1976).

3. The criteria for deciding whether a piece of information qualifies as a MEME, such as *cui bono* ("who benefits"), are discussed by Daniel Dennett (1995). In this connection, it is worth noting that a gene is a kind of meme. Unlike genes, however, memes are both carriers of information and traits (Jablonka and Lamb 2005).

4. With regard to the CALVINIST VIEWS ON HAPPINESS, J. R. Beeke (2004, p. 137) remarks that "self-denial helps us find true happiness." This stance often gets Calvinism a special mention in happiness literature (see, for example, Veenhoven 1994).

5. From the computational standpoint, IMITATION is not as straightforward as it may seem (see Edelman 2008a, ch. 6). The cumulative evolution of tool manufacture in New Caledonian crows has been studied by Gavin Hunt and Russell Gray (2003).

6. For a general introduction to CO-EVOLUTION, see Jablonka and Lamb (2005). The co-evolution of human languages and of *Homo sapiens* is discussed by Morten Christiansen and Nick Chater (2008), whose analysis may be compared to the standard genetics account of language, culture, and intelligence offered by Steven Pinker (2010). Thomas Scott-Phillips and Simon Kirby (2010) describe controlled studies of the role of cultural transmission in language evolution. The functional repercussions of cultural co-evolution in the social and cognitive domains are discussed by Andy Clark (1998).

7. The view of LANGUAGE AS A GAME is due to Wittgenstein (1958).

8. This DIALOGUE between Romeo and Juliet is from act I, scene V. The author of the play almost convinces us that this is how teenagers talk to each other.

9. The ubiquity of SERIAL ORDER IN BEHAVIOR was pointed out famously by Karl Lashley (1951). In vocal articulation, it is speech "gestures" that are combinatorial dynamical action units (Goldstein et al. 2007). In sign language, much use is made of the spatial layout of the extrapersonal space shared by the signer and the spectator, but these signs too are meticulously sequenced. Ann Senghas, Sotaro Kita, and Aslı Özyürek (2004) document the increasingly digital nature of the Nicaraguan sign language as it evolved over successive generations of signers.

10. The importance of LANGUAGE BEING DIGITAL, which is one of the very few of its universal properties (Evans and Levinson 2009), is noted and analyzed in Edelman (2008b, 2008c).

11. Ben-Ami Scharfstein (2009), writing on aesthetics, draws attention to the PLEASURE that people derive from human speech.

12. Douglas Oxley and his colleagues (2008) report a correlation between NOVELTY aversion and conservatism; Michael Cohen and colleagues (2009) relate novelty seeking to individual differences in brain anatomy.

13. Sonja Lyubomirsky, Kennon Sheldon, and David Schkade (2005) discuss the TRANSIENT NATURE OF AFFECT and ways of countering it; see also Edelman (2008a), ch. 10.

14. Indeed, when I first saw it, I did not yet know the scientific name of the SMOKE TREE, or "corona de Christo"; it is *Psorothamnus spinosus*. The snake was a Western shovel-nose, *Chionactis occipitalis*.

15. Michael Anderson (2010) offers a thorough review of REUSE in brain evolution.

16. The RAILROAD SIMILE for language is borrowed from Edelman (2008a), ch. 7.

17. To appreciate fully the fact that AN UTTERANCE MERELY HINTS AT MEANING, consider the following sentence from *Consider Phlebas* (1987), the first space opera in the great Culture series by Iain M. Banks:

> The Jinmoti of Bozlen Two kill the hereditary ritual assassins of the new Yearking's immediate family by drowning them in the tears of the Continental Empathaur in its Sadness Season.

This sentence, which is part of an internal monologue of one of the novel's protagonists, is completely disconnected from its immediate context and is offered no gloss. Yet, a human reader with a reasonably extensive exposure to literature (in particular, historical novels and perhaps science fiction) not only understands what the sentence implies but probably also feels for the king and for his doomed assassins, as well as for the hapless Empathaur. In one deft move, Banks outlines what may be a key characteristic of an entire planetary civilization and makes the reader weigh, even if just for a moment, whether or not being a king is a happy job. The book's title, by the way, is a reference to T. S. Eliot's *The Waste Land*, which brings into the picture a host of connotations by invoking a poem that itself broke all records in this respect when it was first published in 1922. To me, the Bozlen Two customs also connote the Wicker Queen character from Michael Swanwick's outstanding book *The Iron Dragon's Daughter* (1993).

18. The problematic theoretical status of THE MEANING OF "MEANING" is unfortunate: semantics is the one branch of linguistics to which all others report. Equally unfortunate is the custom of dismissing shades of meaning by saying that they are "merely semantic." The argument that this phrase embodies is self-undermining: by revealing the utterers' fundamental lack of care for, or understanding of, what they say, it places under suspicion everything they profess, including the dismissal of semantics.

19. Greg Stephens, Lauren Silbert, and Uri Hasson (2010) report evidence of DYNAMIC COORDINATION between the speaker's and the listener's brain activity patterns.

20. Studies of EMBODIED LANGUAGE PROCESSING (see, for example, Speer et al. 2009) show that listening to stories and experiencing what they describe give rise to similar brain processes.

21. According to Mikhail Bakhtin (quoted in Wertsch 1998), to make an utterance is to "appropriate the words of others and populate them with one's own intention." Edelman (2008a, ch. 7) argues that the knowledge structure that supports this process is a CONSTRUCTION GRAMMAR (Goldberg 2005). Similar structures support behavior in general (Edelman 2011a).

22. Psycholinguistic studies suggest that SEMANTICS IS SUBJECTIVE (see, for example, Stenning and van Lambalgen 2008). There is one and only one situation in which meaning is universal and immutable: inside a closed formal system such as deductive logic or, more generally, mathematics. In all other situations—that is, in all of science and all of what we call "the real world"—things are open to multiple interpretations. We should remember, however, that in a formal system too provability and truth need not coincide, as shown by Gödel (for an entertaining and insightful introduction to these issues, see Hofstadter 1979).

23. WITTGENSTEIN (1958), part I, §525; part I, §534; part II, §VI.

24. Because different people encounter linguistic constructions in somewhat different sets of contexts and may distill out of them somewhat different grammars of experience, one person's "multitude of familiar paths" through language is likely to differ from another's. These INDIVIDUAL DIFFERENCES may be more pronounced on the fringes of humanity's communal body of language—in linguistic games that involve passing around rare or complicated constructions, or in specialized games that are held in various domains of expertise.

25. The logical process of DEDUCTION, which is the foundation of mathematics, consists of proving theorems from premises treated as axioms. Deduction is sound, but cannot generate new knowledge that is not implicit in the premises. In comparison, INDUCTION, which is the basis of all science, generates new knowledge by treating statistical patterns in data as rules that extend to new cases. In that respect, as noted by Hume (1740, part IV, section I), "all knowledge resolves itself into probability."

26. The William James quote is from James (1890), p. 488.

27. Different LANGUAGES carve the space of possibilities differently and employ different sets of constructions; even such seemingly basic categories as noun and verb are not the same across languages (Evans and Levinson 2009). In this respect, the fantasy of Jorge Luis Borges, "Tlön, Uqbar, Orbis Tertius" (1941/1962b, p. 33), does not sound too implausible: "In the languages of the southern hemisphere of Tlön there are no nouns, only verbs; in those of the northern hemisphere, the noun is formed by an accumulation of adjectives." The following trait of another of his inventions is, however, not found in any human language: "There are objects composed of two terms, one of visual and another of auditory character: the colour of the rising sun and the far-away cry of a bird."

Learning the proper referents of words is a problem that in principle is severely underconstrained (Quine 1960). The process through which babies solve it is best characterized as what Charles Sanders Peirce (1868, ms. 692) called "ABDUCTION," or informed guessing: "But we must conquer the truth by guessing, or not at all" (Eco and Sebeok

1988; cf. Sebeok and Sebeok 1981). In a Bayesian formulation, abduction corresponds to evidential reasoning (Edelman 2008a, ch. 8).

28. The WHOLE-OBJECT BIAS in word referent learning was discovered by Ellen Markman (1989). Infants tend to associate a new verbal label first with the most salient novel shape present, then with texture and color.

29. Regarding the role of DEPARTURE FROM EQUIPROBABILITY, Zellig Harris (1991, p. 32) writes: "It is an essential property of language that the combinations of words and utterances are not all equiprobable. It follows that whatever else there is to be said about the form of language, a fundamental task is to state the departures from equiprobability in sound- and word-sequences."

30. In Fitzgerald's English translation of *The Odyssey*, a different word ("long") is repeated in this passage, which masterfully approximates the effect of the repetition in the original Greek.

31. Phonemes too can be discovered by the same ALIGN-AND-COMPARE procedure (Edelman 2008a, ch. 7). For a sample of empirical studies of this issue, see Onnis, Waterfall, and Edelman (2008), and Saffran, Aslin, and Newport (1996); Saffran and Wilson (2003); for a general theoretical framework, see Goldstein et al. (2010).

32. VARIATION SETS appear in child-directed speech with approximately the same frequency in all languages that have been examined (Waterfall and Edelman 2009). See Waterfall et al. (2010) for a detailed study of variation sets in English, and Goldstein et al. (2010) for a theoretical discussion.

33. INDIVIDUAL DIFFERENCES in comprehension are widespread and profound (Chipere 2001; Dabrowska and Street 2006); they depend, among other factors, on the socioeconomic background of the subject (Hackman, Farah, and Meaney 2010; Hoff 2003). It is becoming increasingly clear that parents' depression affects caregiving, which in turn affects children's language development (Stein et al. 2008), in particular vocabulary (Paulson, Keefe, and Leiferman 2009). The nuances of timing and social feedback that mediate those effects are beginning to be understood (Goldstein et al. 2010). It should be noted that children's language abilities at age five predict life outcomes decades later (Schoon et al. 2010).

34. The DIALOGUE is from the Santa Barbara Corpus of spoken English (SBC015: 849.456–856.005); it is quoted from Du Bois (forthcoming).

35. The imaging study that demonstrated brain COORDINATION between speakers and listeners is Stephens et al. (2010).

NOTES TO CHAPTER 6

1. In Buddhist philosophy, the technical term for the concept that I chose to render as "the web of cause and effect" is INTERDEPENDENT CO-ORIGINATION—the principle according to which nothing in the mental domain (which includes the mind's representation of the world, such as it is) is uncaused or is without origin. Nāgārjuna extends this to the universe in general, thus arriving at the emptiness doctrine, which states that nothing has existence in itself; rather, everything is defined by its place in the web of cause and effect (Gier and Kjellberg 2004).

2. Regarding BRAIN DYNAMICS, Lashley (1951, p. 153) writes: "I can best illustrate this conception of nervous action by picturing the brain as the surface of a lake." The dynamics of disturbances in a body of water can be used, by the way, to form a spatial representation of its layout (Buckingham, Potter, and Epifanio 1996).

3. The best expression of the world's predictability in this respect is David Marr's (1970, pp. 150–151) FUNDAMENTAL HYPOTHESIS: "Where instances of a particular collection of intrinsic properties (i.e., properties already diagnosed from sensory information) tend to be grouped such that if some are present, most are, then other useful properties are likely to exist which generalize over such instances. Further, properties often are grouped in this way." This can be compared to "the single most important principle underlying the mechanisms of perception and conscious experience: that they may have evolved exclusively for extracting statistical regularities from the natural world" (Ramachandran and Hirstein 1997, p. 453). It should be noted that humans are capable of learning such regularities from just two or three repetitions (Turk-Browne et al. 2009).

4. The analysis of the perception of STRAIGHTNESS follows O'Regan and Noë (2001).

5. The CURVATURE ADAPTATION experiments are from Gibson (1933).

6. For a review of the embodiment thesis in cognitive science, see Anderson (2003). The examples involving emotion are from Niedenthal et al. (2005). The realization that COGNITION IS EMBODIED AND SITUATED prompts researchers such as Dennis Proffitt (2006) to quote Protagoras: "Man is the measure of all things."

7. Edelman (1999) documents VERIDICAL PERCEPTION of shape and offers a computational analysis of the conditions under which it obtains.

8. The line is from John Donne's "Epithalamion Made at Lincoln Inn" (1595). Ellen Berscheid (2010, p. 15), in her taxonomic review of the psychology of LOVE, cites Aron and Aron (1986), who "believe that certain rapid changes in a new relationship, namely, the rapid 'expansion of the self' or incorporation into the self-concept of the qualities of the other, produce the euphoria often associated with falling in love." The role of the Self in love is illuminated by this story from the writings of the Sufi poet and mystic Mowlana Jalal ad-Din Rumi (1207–1273):

> One went to the door of the Beloved and knocked. A voice asked: "Who is there?" He answered: "It is I." The voice said: "There is no room here for me and thee." The door was shut. After a year of solitude and deprivation, this man returned to the door of the Beloved. He knocked. A voice from within asked: "Who is there?" The man said: "It is Thou." The door was opened for him.

A version of Rumi's story appears, in verse, in *The Mesnavi and the Acts of the Adepts* (Rumi and Ahmed 1881), p. 221.

9. CAUSAL INFERENCE plays a key role in perception (Shams and Beierholm 2010) and in reasoning (Gopnik et al. 2004). An analysis of intervention as Bayesian "explaining away" can be found in Steyvers et al. (2003). Tamar Kushnir and Alison Gopnik (2005) found that children trust their own interventions much more than observation. People take this difference into account when making inferences from contingency to agency (Moore et al. 2009), as do rats (Leising et al. 2008). There is some evidence of causal and analogical reasoning (that is, transfer to new situations) in New Caledonian crows (Taylor et al. 2009).

10. Joshua Greene and Jonathan Cohen (2004) offer an introduction to CONSEQUENTIALIST ETHICS; see also Edelman (2008a, ch. 10).

11. From Shakespeare's *A Midsummer Night's Dream*, act V, scene I, lines 19–22.

12. Deviations from routine, which constitutes a large proportion of human activity (Eagle and Pentland 2009; Song et al. 2010), are relevant in INFERRING AGENCY (Auvray, Lenay, and Stewart 2009; Waytz et al. 2010). The evolution of agency in the deterministic universe is explained by Dennett (2003a).

13. Daniel Wegner (2004, p. 688) explains that "the whole idea of a 'PERSON' is an elegant accounting system for making sense of actions and ascribing them to constructed entities that are useful for purposes of social justice and the facilitation of social interaction." From the computational standpoint, the "person" construct is seen to facilitate learning (Glymour 2004).

14. Convergence between intrapersonal PROSPECTION (Roberts and Feeney 2009) and interpersonal PERSPECTIVE-TAKING suggests that the same causes of mind perception toward others might drive mind perception toward one's future self (Waytz et al. 2010). As a GENERATIVE predictive model, a representation of the Self can be implemented by a mechanism that is widely applicable across cognition and therefore amenable to evolutionary reuse (Anderson 2010).

15. The categorization of *Homo sapiens* as a GENERALIST species is advanced by Dobzhansky (1972).

16. Positive affect is known to be an implicit motivator (Custers and Aarts 2005). POSITIVE MOOD broadens associations and facilitates creativity (Bar 2009; Isen 2001). The implications of the differences between modern and ancestral environments for the motivation for, and the ease of, learning are discussed in Geary (2009).

17. Mihaly Csikszentmihalyi and Jeremy Hunter (2003) describe a typical application of mood sampling to the study of FLOW and happiness (cf. Collins, Sarkisian, and Winner 2009). People's enjoyment of overcoming obstacles has been studied, for instance, by Labroo and Kim (2009).

18. In the Canadian film *One Week*, written and directed by Michael McGowan, the protagonist, Ben Tyler, asks a woman he met on a trail, hiking in the Banff National Park, "What would you be doing if you had one week to live?" She replies, "I'd be doing what I am doing NOW."

19. For reviews of the benefits of POSITIVE AFFECT for decision making and problem solving, see Isen (2001, 2008). These benefits are not indiscriminate; rather, "positive affect enables flexible thinking about topics that people want or have to think about" (Ashby, Isen, and Turken 1999, p. 531). Furthermore, "for a complete theory of positive affect, it is necessary to understand why certain things make people happy, even if it were known that dopamine is released when people are happy, and why dopamine release has the particular consequences it does on cognition." Interestingly, better problem-solving abilities correlate with less depression, even in octogenarians (Margrett et al. 2010).

20. The effects of positive mood on CREATIVITY are reviewed by Barbara Fredrickson (1998). Ed Diener, Richard Lucas, and Christie Napa Scollon (2006, p. 307), noting that positive moods facilitate a variety of approach behaviors and positive outcomes, conclude: "Thus, the ubiquity of a positive emotional set point, in concert with the less frequent experience of unpleasant emotions, likely results from the ADAPTIVE NATURE OF FREQUENT POSITIVE EMOTIONS."

21. Possible anatomical links between the brain circuits involved in HEDONIA AND EUDAIMONIA and connections to the so-called default network (Mason et al. 2007), are listed in Kringelbach and Berridge (2009).

22. For a view of the narrative Self as the "CENTER OF NARRATIVE GRAVITY," see Dennett (1991). The collection titled *The Mind's I* (Hofstadter and Dennett 1981) contains some fascinating stories that shed light on many aspects of selves, including their physical location.

23. Readers who feel that they have finally had enough of my repeated appeals to FICTIONAL CHARACTERS may wish to consult Garfield (2006), who wrote that "fictions can constitute worlds against which truth can be assessed, despite the fact that those worlds are themselves fictional." As to how the thoughts of the dead can affect the living, see Edelman (2008a, pp. 444–447).

24. Dan Lloyd's "Music of the Hemispheres," http://indexmagazine .com/vid-music_of_hemispheres.html, describes a practical approach to turning brain dynamics into actual MUSIC.

25. Bloom's piece, titled "First Person Plural," appeared in the *Atlantic* in November 2008.

26. Thomas Nagel's celebrated paper "What Is It Like to Be a Bat?" (1974) marks the origin of this kind of philosophical analysis. Nagel's claim that the answer to his question is unknowable (owing to the nonexistence of the mode of knowledge that Dennett calls HETEROPHE-NOMENOLOGY) has been disputed; see Dennett (1991, 2003b) and Clark (2000, p.129). Practical aspects of these issues as they apply in medicine are discussed in Merker (2007). For a computational treatment by a *very* informed science fiction author, see the chapter titled "Wang's Carpets" in Greg Egan's novel *Diaspora* (1997).

27. The identification of experience with a TRAJECTORY through a state space is introduced by Spivey (2006) and defended and analyzed in great detail by Fekete and Edelman (2011).

28. The notion of JUST SEEING, as distinguished from Wittgenstein's SEEING AS, is discussed in Edelman (2009); see also Fekete and Edelman (2011).

29. In real life, the ZOMBIE state is approximated by the pathological condition known as absence seizures (Metzinger 2003).

30. For an introduction to the functional neuroanatomy of VERTEBRATE-LIKE EXPERIENCE, which requires a very particular kind of neural dynamics, see Merker (2007). For a fully human-like experience, a human-like cortex is needed (Merker 2004). The four functional ingredients of the PHENOMENAL SELF are from Metzinger (2003). Simple ways to induce an out-of-body experience in normal subjects have been explored by Blanke and Metzinger (2009).

31. The logical absurdity and psychological vapidity of the classical notion of FREE WILL have been pointed out by Hume (1740) and by Voltaire (1752/1924). See Wegner (2004) for a modern psychological stance; Gier and Kjellberg (2004) for a comparative-philosophical perspective; and Edelman (2008a, ch. 10) for an integrated treatment.

32. Because it is not possible to DISTINGUISH DREAMING FROM REAL-ITY (except on statistical grounds by means that are necessarily fallible because the baseline statistics can be fantasized or subverted by an evil demon in charge of the simulated reality), the rational course of action is to take reality at face value (Edelman 2011b). This implies that even a first-person phenomenal insight into the nature of reality has no

practical consequences: you can be dead sure that this is all just a dream, but this intuitive certainty too can be an illusion. However, to adhere to this principle, one has either to remember it or to be able to derive it, which may not be possible in various states of reduced cognitive function, including dreaming, or being simulated on inferior hardware. You may have suspected all along that enlightenment is overrated. Now you know why.

33. Ruess to Waldo Ruess, November 11, 1934, in W. L. Rusho, *Everett Ruess: A Vagabond for Beauty* (Salt Lake City: Peregrine Smith Books, 1983), pp. 179–180.

NOTES TO CHAPTER 7

1. An ongoing study, involving thousands of people, that uses an iPhone application to sample phone owners' patterns of MIND WANDERING and subjective well-being (Killingsworth and Gilbert 2010) has yielded some results that bear on the "Prometheus on parole" scenario. First, people's minds appear to wander frequently, regardless of the activity they are engaged in. Second, people are less happy when their minds are wandering than when they are not (compare this with the idea of "flow" discussed in Chapter 6). The negative effect, however, is due to a large extent to bouts of thinking about unpleasant things, which suggests, unsurprisingly, that while traveling mentally you should avoid visiting nasty places. I might add that if you *are* in a place "where but to think is to be full of sorrow" (as Keats put it in "Ode to a Nightingale") to begin with, then any travel would seem to be good for you.

2. This statement, which according to Dave Barry contains his entire PHILOSOPHY OF LIFE, is found on the very first page of *Dave Barry's Bad Habits* (Barry 1993).

3. This answer, given by Shimon Ben Zoma to the question "WHOSOEVER IS RICH?," is recorded in the Talmud (Mishna, Avot 4).

4. Lao Tze (1904).

5. The historical meeting between ALEXANDER AND DIOGENES in Corinth is mentioned by Jorge Luis Borges in a 1953 essay, "The Dia-

logues of Ascetic and King" (1953/1999, p. 382). I am indebted to Björn Merker for sharing with me parts of his unpublished manuscript "Sapta Svapna Sutra" ("Seven Dreams Dialogue"), in which the choice between world conquest and self-conquest is framed by an imaginary debate between the Buddha and Genghis Khan.

6. The optimality of having one's desires coincide with one's means is backed empirically by a recent paper whose title is "Is Happiness Having What You Want, Wanting What You Have, or Both?" and whose conclusion is: "Both" (Larsen and McKibban 2008, p. 371).

7. The value of SELF-KNOWLEDGE is a central tenet of all three great philosophical traditions: Greek (Aristotle 350 B.C.E.; Nagel 1972) Chinese (Tze 1904, ch. 33), and Indian (Eliot 1921, p. 475).

8. The SAUSAGE simile is adapted from Birdsong (1989, p. 25), who decried the reliability of subjective judgments of grammaticality by comparing them to hot dogs.

9. The Royal Society's motto, NULLIUS IN VERBA, is a contraction of the following couplet from Horace's Epistles, Book I.I, lines 13–14:

Ac ne forte roges, quo me duce, quo lare tuter,

Nullius addictus iurare in verba magistri.

(Lest by chance you ask who leads me, by which household god I am sheltered, I swear by the words of no master.)

10. The lines beginning with "NOW UNDERSTAND ME WELL . . ." by Walt Whitman are quoted from Leaves of Grass: Song of the Open Road (1892/1990, 82:14).

11. I thank Björn Merker for suggesting to me that a past history of consistent rewards pushes people to persevere in a course of action that may have exhausted its use. The appetite for NOVELTY and the construal and appreciation of reward are, of course, expected to vary widely between individuals, as attested by many studies. One of the most fascinating findings in this domain is that of Chen, Burton, Greenberger, and Dmitrieva (1999), who looked for possible genetic factors contributing to long-range migrations, such as the millennia-long trek that eventually brought humans out of Africa to the tip of South America. This study revealed a strong correlation between the population's locus along the migration arc and the preponderance in

that population of long alleles (for example, 7-repeats) of the DRD4 gene—a gene that codes for a dopamine receptor that has been linked to novelty-seeking personality, as well as to hyperactivity and risk-taking behaviors.

12. Concerning the challenge of SELF-CONQUEST, the Talmud (Mishna, Avot 4) contains the following exchange: "Whosoever is mighty? He who conquers his passions, as it is written (Proverbs 16:32) 'One who is slow to anger is better than the mighty, and one whose temper is controlled than one who captures a city.'" In the Muslim tradition, self-conquest is called the "greater jihad" or *jihad al-nafs* (the struggle against the soul) and is usually elevated over the "lesser jihad" or holy war; cf. Ibn al-Jawz (1998, p. 122): "I reflected over jihad al-nafs and realised it to be the greatest jihad."

13. For the first computationally inspired exploration of the idea of THE SOCIETY OF MIND, see Minsky (1985).

14. The demonstrator killed by the police on June 2, 1967, was BENNO OHNESORG; the officer who shot him was later identified as an agent of the East German secret police, the Stasi (Kulish 2009).

15. The abolitionist JOHN BROWN was hanged by the State of Virginia on December 2, 1859.

16. For an illuminating discussion of the roots of PRACTICAL-WISDOM humanism in Eastern and Western thinking, see Gier (2002).

17. Concerning PRACTICAL WISDOM, Aristotle wrote in *Nicomachean Ethics* (350 B.C.E., 1142a): "Whereas young people become accomplished in geometry and mathematics, and wise within these limits, prudent young people do not seem to be found. The reason is that prudence is concerned with particulars as well as universals, and particulars become known from experience, but a young person lacks experience, since some length of time is needed to produce it."

18. The concept of a NARRATIVE SELF, which I discussed in an earlier chapter, can be summarized by the following passage from the afterword to *The Maker* by Jorge Luis Borges (1964): "A man sets out to draw the world. As the years go by, he peoples a space with images of provinces, kingdoms, mountains, bays, ships, islands, fishes, rooms, instruments, stars, horses, and individuals. A short time before he dies,

he discovers that the patient labyrinth of lines traces the lineaments of his own face."

19. The tendency of phenomenal experience, via the memories it creates, to enrich the appreciation of further experience is the reason why, hedonically speaking, THE RICH GET RICHER. It is probably also why older folks report routine levels of well-being that surprise younger researchers (Stone et al. 2010). Interestingly, material riches don't have quite the same effect: Daniel Kahneman and Angus Deaton (2010) report that high income improves evaluation of life but not emotional well-being.

20. What is UBIK?

> I am Ubik. Before the universe was, I am. . . . I am the word and my name is never spoken, the name which no one knows. I am called Ubik, but that is not my name. I am. I shall always be.
>
> —PHILIP K. DICK, *Ubik* (1969)

FURTHER READING

Adelson, E. H., and A. P. Pentland. 1996. "The Perception of Shading and Reflectance." In *Perception as Bayesian Inference*, ed. D. Knill and W. Richards, pp. 409–423. New York: Cambridge University Press.

Al-Jawz, Ibn. 1998. *Sayd al-Khatir*. al-Mansurah, Egypt: Dar al-Yaqin.

Alighieri, Dante. 1982. *The Divine Comedy*, trans. Geoffrey L. Bickersteth. Oxford: Basil Blackwell.

Anderson, M. L. 2003. "Embodied Cognition: A Field Guide." *Artificial Intelligence* 149: 91–130.

———. 2010. "Neural Re-use as a Fundamental Organizational Principle of the Brain." *Behavioral and Brain Sciences* 33(4): 245–313.

Aristotle. 350 B.C.E. *Nicomachean Ethics*, trans. W. D. Ross. Available at: http://classics.mit.edu/Aristotle/nicomachaen.html.

Aron, A., and E. N. Aron. 1986. *Love and the Expansion of Self: Understanding Attraction and Satisfaction*. New York: Hemisphere.

Ashby, F. G., A. M. Isen, and A. U. Turken. 1999. "A Neuropsychological Theory of Positive Affect and Its Influence on Cognition." *Psychological Review* 106: 529–550.

Auvray, M., C. Lenay, and J. Stewart. 2009. "Perceptual Interactions in a Minimalist Virtual Environment." *New Ideas in Psychology* 27: 32–47.

Avni, R., Y. Tzvaigrach, and D. Eilam. 2008. "Exploration and Navigation in the Blind Mole Rat (*Spalax ehrenbergi*): Global Calibration as a Primer of Spatial Representation." *Journal of Experimental Biology* 211: 2817–2826.

Baer, R. A. 2003. "Mindfulness Training as a Clinical Intervention: A Conceptual and Empirical Review." *Clinical Psychology: Science and Practice* 10: 125–143.

Bailey, D. H., and D. C. Geary. 2009. "Hominid Brain Evolution: Testing Climatic, Ecological, and Social Competition Models." *Human Nature* 20: 265–279.

Banks, I. M. 1987. *Consider Phlebas*. London: Macmillan.

Bar, M. 2009. "A Cognitive Neuroscience Hypothesis of Mood and Depression." *Trends in Cognitive Sciences* 13: 456–463.

Barash, D. P., P. Donovan, and R. Myrick. 1975. "Clam Dropping Behavior of the Glaucous-Winged Gull (*Larus glaucescens*)." *Wilson Bulletin* 87: 60–64.

Barry, D. 1993. *Dave Barry's Bad Habits*. New York: Macmillan.

Beeke, J. R. 2004. "Calvin on Piety." In *The Cambridge Companion to John Calvin*, ed. Donald K. McKim, pp. 125–151. Cambridge: Cambridge University Press.

Berscheid, E. 2010. "Love in the Fourth Dimension." *Annual Review of Psychology* 61: 1–25.

Birdsong, D. 1989. *Metalinguistic Performance and Interlinguistic Competence*. New York: Springer.

Bishop, C. M. 2006. *Pattern Recognition and Machine Learning*. Berlin: Springer-Verlag.

Blanke, O., and T. Metzinger. 2009. "Full-Body Illusions and Minimal Phenomenal Selfhood." *Trends in Cognitive Sciences* 13: 7–13.

Bloom, P. 2008. "First Person Plural." *Atlantic*, November.

Borges, J. L. 1960. *The Maker*, trans. Mildred Boyer. In *Dreamtigers* (*El Hacedor*), trans. M. Boyer and H. Morland. Austin: University of Texas Press. Originally published in 1940.

———. 1962a. *The Approach to Al-Mu'tasim* by Mir Bahadur Ali, trans. A. Bonner in collaboration with the author. In *Ficciones*. New York: Grove Press. Originally published in 1935.

———. 1962b. "The Garden of the Forking Paths," "Tlön, Uqbar, Orbis Tertius," trans. A. Bonner in collaboration with the author. In *Ficciones*. New York: Grove Press. Originally published in 1941.

———. 1999. "The Dialogues of Ascetic and King." In *Selected Non-Fictions*, ed. E. Weiberger. New York: Viking. Originally published in 1953.

———. 2000. "Happiness." In *Jorge Luis Borges: Selected Poems*, ed. Alexander Coleman. New York: Penguin.

Buckingham, M. J., J. R. Potter, and C. L. Epifanio. 1996. "Seeing in the Ocean with Background Noise." *Scientific American* 274: 86–90.

Bullitt, J. T., ed. "Tipitaka: The Pali Canon." *Access to Insight*, 10 May 2011, http://www.accesstoinsight.org/tipitaka/index.html.

Caporale, N., and Y. Dan. 2008. "Spike Timing-Dependent Plasticity: A Hebbian Learning Rule." *Annual Review of Neuroscience* 31: 25–46.

Carroll, L. 1865. *Alice's Adventures in Wonderland*. London: Macmillan.

Chalmers, D. J. 1994. "On Implementing a Computation." *Minds and Machines* 4: 391–402.

Chater, N., J. B. Tenenbaum, and A. Yuille. 2006. "Probabilistic Models of Cognition: Conceptual Foundations." *Trends in Cognitive Sciences* 10: 287–291.

Chen, C., M. Burton, E. Greenberger, and J. Dmitrieva. 1999. "Population Migration and the Variation of Dopamine D4 Receptor (DRD4) Allele Frequencies Around the Globe." *Evolution and Human Behavior* 20: 309–324.

Chengen, Wu. 2006. *The Monkey and the Monk: An Abridgment of* The Journey to the West, trans. A. C. Yu. Chicago: University of Chicago Press.

Chipere, N. 2001. "Native Speaker Variations in Syntactic Competence: Implications for First Language Teaching." *Language Awareness* 10: 107–124.

Christiansen, M. H., and N. Chater. 2008. "Language as Shaped by the Brain." *Behavioral and Brain Sciences* 31: 489–509.

Clark, A. 1998. "Magic Words: How Language Augments Human Computation." In *Language and Thought: Interdisciplinary Themes*, ed. P. Carruthers and J. Boucher, pp. 162–183. Cambridge: Cambridge University Press.

Clark, A. 2000. *A Theory of Sentience*. Oxford: Oxford University Press.

Clayton, P., and S. A. Kauffman. 2006. "Agency, Emergence, and Organization." *Biology and Philosophy* 21: 501–521.

Cohen, M. X., J.-C. Schoene-Bake, C. E. Elger, and B. Weber. 2009. "Connectivity-Based Segregation of the Human Striatum Predicts Personality Characteristics." *Nature Neuroscience* 12: 32–34.

Collins, A. L., N. Sarkisian, and E. Winner. 2009. "Flow and Happiness in Later Life: An Investigation into the Role of Daily and Weekly Flow Experiences." *Journal of Happiness Studies* 10: 703–719.

Colombo, M., and N. Broadbent. 2000. "Is the Avian Hippocampus a Functional Homologue of the Mammalian Hippocampus?" *Neuroscience and Biobehavioral Reviews* 24: 465–484.

Crick, F. 1994. *The Astonishing Hypothesis: The Scientific Search for the Soul.* New York: Charles Scribner's Sons.

Csikszentmihalyi, M., and J. Hunter. 2003. "Happiness in Everyday Life: The Uses of Experience Sampling." *Journal of Happiness Studies* 4: 185–199.

Custers, R., and H. Aarts. 2005. "Positive Affect as Implicit Motivator: On the Nonconscious Operation of Behavioral Goals." *Journal of Personality and Social Psychology* 89: 129–142.

Dabrowska, E., and J. Street. 2006. "Individual Differences in Language Attainment: Comprehension of Passive Sentences by Native and Non-native English Speakers." *Language Sciences* 28: 604–615.

Daniel, T. C. 1990. "Measuring the Quality of the Natural Environment: A Psychophysical Approach." *American Psychologist* 45: 633–637.

Davidson, R. J. 2010. "Empirical Explorations of Mindfulness: Conceptual and Methodological Conundrums." *Emotion* 10: 8–11.

Davis, S., B. Bozon, and S. Laroche. 2003. "How Necessary Is the Activation of the Immediate Early Gene zif268 in Synaptic Plasticity and Learning?" *Behavioral Brain Research* 142: 17–30.

Dawkins, R. 1976. *The Selfish Gene.* Oxford: Oxford University Press.

Deisseroth, K., P. G. Mermelstein, H. Xia, and R. W. Tsien. 2003. "Signaling from Synapse to Nucleus: The Logic Behind the Mechanisms." *Current Opinion in Neurobiology* 13: 354–365.

DeLoache, J. S., and V. LoBue. 2009. "The Narrow Fellow in the Grass: Human Infants Associate Snakes and Fear." *Developmental Science* 12: 201–207.

Dennett, D. C. 1987. *The Intentional Stance.* Cambridge, MA: MIT Press.

———. 1991. *Consciousness Explained.* Boston: Little, Brown.

———. 1995. *Darwin's Dangerous Idea: Evolution and the Meanings of Life.* New York: Simon & Schuster.

———. 2003a. *Freedom Evolves.* New York: Viking.

———. 2003b. "Who's on First? Heterophenomenology Explained." *Journal of Consciousness Studies* 10: 19–30.

Diamond, R., and S. Carey. 1986. "Why Faces Are and Are Not Special: An Effect of Expertise." *Journal of Experimental Psychology* 115(2): 107–117.

Dick, P. K. 1969. *Ubik*. New York: Doubleday.

———. 1993. "The Short Happy Life of a Brown Oxford." In *The Collected Stories of Philip K. Dick*. New York: Carol Publishing Group. Originally published in *Fantasy and Science Fiction*, January 1954.

Diener, E., R. E. Lucas, and C. Napa Scollon. 2006. "Beyond the Hedonic Treadmill: Revising the Adaptation Theory of Well-being." *American Psychologist* 61: 305–314.

Dobzhansky, T. 1972. "The Ascent of Man." *Social Biology* 19: 367–378.

———. 1973. "Nothing in Biology Makes Sense Except in the Light of Evolution." *The American Biology Teacher* 35: 125–129.

Donne, J. 1996. "Epithalamion made at Lincoln Inn." In *Selected Poetry*, ed. John Carey. Oxford: Oxford University Press.

Du Bois, J. W. Forthcoming. "Towards a Dialogic Syntax."

Eagle, N., and A. Pentland. 2009. "Eigenbehaviors: Identifying Structure in Routine." *Behavioral Ecology and Sociobiology* 63: 1057–1066.

Eco, U., and T. A. Sebeok. 1988. *The Sign of Three: Dupin, Holmes, Peirce*. Bloomington: Indiana University Press.

Edelman, S. 1998. "Spanning the Face Space." *Journal of Biological Systems* 6: 265–280.

———. 1999. *Representation and Recognition in Vision*. Cambridge, MA: MIT Press.

———. 2006. "Mostly Harmless: Review of *Action in Perception* by Alva Noë." *Artificial Life* 12: 183–186.

———. 2008a. *Computing the Mind: How the Mind Really Works*. New York: Oxford University Press.

———. 2008b. "On the Nature of Minds; or, Truth and Consequences." *Journal of Experimental and Theoretical AI* 20: 181–196.

———. 2008c. "A Swan, a Pike, and a Crawfish Walk into a Bar." *Journal of Experimental and Theoretical AI* 20: 261–268.

———. 2009. "On What It Means to See, and What We Can Do About It." In *Object Categorization: Computer and Human Vision Perspectives*, ed. S. Dickinson, A. Leonardis, B. Schiele, and M. J. Tarr, pp. 69–86. Cambridge: Cambridge University Press.

———. 2011a. "On Look-Ahead in Language: Navigating a Multitude of Familiar Paths." In *Predictions in the Brain*, ed. M. Bar, pp. 170–189. New York: Oxford University Press.

———. 2011b. "Regarding Reality: Some Consequences of Two Incapacities." *Frontiers in Theoretical and Philosophical Psychology*, volume 2, article 44.

———. 2012. *The Happiness of Pursuit*. New York: Basic Books.

Edelman, S., and S. Duvdevani-Bar. 1997. "Similarity-Based Viewspace Interpolation and the Categorization of 3D Objects." In *Proceedings of the Similarity and Categorization Workshop* (University of Edinburgh, Department of Artificial Intelligence), pp. 75–81.

Edelman, S., D. Reisfeld, and Y. Yeshurun. 1992. "Learning to Recognize Faces from Examples." In *Proceedings of the Second European Conference on Computer Vision: Lecture Notes in Computer Science*, ed. G. Sandini, vol. 588, pp. 787–791. London: Springer-Verlag.

Egan, G. 1994. *Permutation City*. London: Orion.

———. 1997. *Diaspora*. New York: HarperCollins.

Eifuku, S., W. C. De Souza, R. Tamura, H. Nishijo, and T. Ono. 2004. "Neuronal Correlates of Face Identification in the Monkey Anterior Temporal Cortical Areas." *Journal of Neurophysiology* 91: 358–371.

Eliot, C. 1921. *Hinduism and Buddhism: An Historical Sketch*. London: Routledge & Kegan Paul.

Eliot, T. S. 1964. "The Hollow Men" and "The Wasteland." In *Collected Poems*. London: Faber and Faber.

———. 1968. "Little Gidding," V. In *Four Quartets*. New York: Mariner Books. Originally published in 1942.

Endres, R. G., and N. S. Wingreen. 2008. "Accuracy of Direct Gradient Sensing by Single Cells." *Proceedings of the National Academy of Sciences* 105: 15749–15754.

Evans, N., and S. Levinson. 2009. "The Myth of Language Universals: Language Diversity and Its Importance for Cognitive Science." *Behavioral and Brain Sciences* 32: 429–492.

Fekete, T., and S. Edelman. 2011. "Towards a Computational Theory of Experience." *Consciousness and Cognition* 20: 807–827.

Ferguson, M. J., and J. A. Bargh. 2004. "Liking Is for Doing: The Effects of Goal Pursuit on Automatic Evaluation." *Journal of Personality and Social Psychology* 87: 557–572.

Feuchtwanger, Lion. 1963. *Der Jüdischer Krieg* ("The Jewish War," 1932), translated into Russian by Vera Stanevich. Moscow: State Publishing House.

Flannelly, K. J., H. G. Koenig, C. G. Ellison, K. Galek, and N. Krause. 2006. "Belief in Life After Death and Mental Health: Findings from a National Survey." *Journal of Nervous and Mental Disease* 194: 524–529.

Fredrickson, B. L. 1998. "What Good Are Positive Emotions?" *Review of General Psychology* 2: 300–319.

Garfield, J. L. 2006. "Reductionism and Fictionalism: Comments on Siderits." *APA Newsletter on Asian and Comparative Philosophy* 6, no. 1: 1–8.

Garfield, J. L., and G. Priest. 2009. "Mountains Are Just Mountains." In *Pointing at the Moon: Buddhism, Logic, Analysis*, ed. M. D'Amato, J. L. Garfield, and T. Tillemans, pp. 71–82. New York: Oxford University Press.

Geary, D. C. 2009. "The Why of Learning." *Educational Psychologist* 44: 198–201.

Gibson, J. J. 1933. "Adaptation, After-effect, and Contrast in the Perception of Curved Lines." *Journal of Experimental Psychology* 16: 1–31.

———. 1979. *The Ecological Approach to Visual Perception*. Boston: Houghton Mifflin.

Gier, N. F. 2000. *Spiritual Titanism: Indian, Chinese, and Western Perspectives*. Albany: State University of New York Press.

———. 2002. "The Virtues of Asian Humanism." *Journal of Oriental Studies* 12: 14–28.

———. 2007. "Hebrew and Buddhist Selves: A Constructive Postmodern Proposal." *Asian Philosophy* 17: 47–64.

Gier, N., and P. K. Kjellberg. 2004. "Buddhism and the Freedom of the Will." In *Freedom and Determinism: Topics in Contemporary Philosophy*, ed. J. K. Campbell, D. Shier, and M. O'Rourke, pp. 277–304. Cambridge, MA: MIT Press.

Gilbert, D. 2006. *Stumbling on Happiness*. New York: Knopf.

Glasheen, J. W., and T. A. McMahon. 1996. "Size-Dependence of Water-Running Ability in Basilisk Lizards (*Basiliscus basiliscus*)." *Journal of Experimental Biology* 199: 2611–2618.

Glimcher, P. W., C. Camerer, R. A. Poldrack, and E. Fehr. 2008. *Neuroeconomics: Decision Making and the Brain*. New York: Academic Press.

Glymour, C. 2001. *The Mind's Arrows*. Cambridge, MA: MIT Press.

———. 2004. "We Believe in Freedom of the Will so That We Can Learn." *Behavioral and Brain Sciences* 27: 661–662.

Gold, J. I., and M. N. Shadlen. 2002. "Banburismus and the Brain: Decoding the Relationship Between Sensory Stimuli, Decisions, and Reward." *Neuron* 36: 299–308.

Goldberg, A. E. 2005. *Constructions at Work: The Nature of Generalization in Language*. Oxford: Oxford University Press.

Goldstein, L., M. Pouplier, L. Chen, E. Saltzman, and D. Byrd. 2007. "Dynamic Action Units Slip in Speech Production Errors." *Cognition* 103: 386–412.

Goldstein, M. H., H. R. Waterfall, A. Lotem, J. Halpern, J. Schwade, L. Onnis, and S. Edelman. 2010. "General Cognitive Principles for Learning Structure in Time and Space." *Trends in Cognitive Sciences* 14: 249–258.

Goodson, J. L. 2005. "The Vertebrate Social Behavior Network: Evolutionary Themes and Variations." *Hormones and Behavior* 48: 11–22.

Gopnik, A., C. Glymour, D. M. Sobel, L. E. Schulz, T. Kushnir, and D. Danks. 2004. "A Theory of Causal Learning in Children: Causal Maps and Bayes Nets." *Psychological Review* 111: 3–32.

Greeley, A. M., and M. Hout. 1999. "Americans' Increasing Belief in Life After Death: Religious Competition and Acculturation." *American Sociological Review* 64: 813–835.

Greene, J. D., and J. D. Cohen. 2004. "For the Law, Neuroscience Changes Nothing and Everything." *Philosophical Transactions of the Royal Society of London B* 359: 1775–1785.

Grodzinski, U., and N. S. Clayton. 2010. "Problems Faced by Food-Caching Corvids and the Evolution of Cognitive Solutions." *Philo-*

sophical Transactions of the Royal Society B 365, no. 1542: 977–987.

Hackman, D. A., M. J. Farah, and M. J. Meaney. 2010. "Socioeconomic Status and the Brain: Mechanistic Insights from Human and Animal Research." *Nature Reviews Neuroscience* 11: 651–659.

Han, K.-T. 2007. "Responses to Six Major Terrestrial Biomes in Terms of Scenic Beauty, Preference, and Restorativeness." *Environment and Behavior* 39: 529–556.

Hardie, P. R. 1985. "Imago Mundi: Cosmological and Ideological Aspects of the Shield of Achilles." *Journal of Hellenic Studies* 105: 11–31.

Harris, Z. S. 1991. *A Theory of Language and Information*. Oxford: Clarendon Press.

Hassabis, D., D. Kumaran, S. D. Vann, and E. A. Maguire. 2007. "Patients with Hippocampal Amnesia Cannot Imagine New Experiences." *Proceedings of the National Academy of Sciences* 104: 1726–1731.

Haxby, J. V., M. I. Gobbini, M. L. Furey, A. Ishai, J. L. Schouten, and P. Pietrini. 2001. "Distributed and Overlapping Representations of Faces and Objects in Ventral Temporal Cortex." *Science* 293: 2425–2430.

Hebb, D. 1949. *The Organization of Behavior*. New York: John Wiley & Sons.

Hesse, Herman. 1969. *The Glass Bead Game (Magister Ludi)*. New York: Holt, Rinehart and Winston.

Hockema, S. A., and L. B. Smith. 2009. "Learning Your Language, Outside-In and Inside-Out." *Linguistics* 47: 453–479.

Hoff, E. 2003. "Causes and Consequences of SES-Related Differences in Parent-to-Child Speech." In *Socioeconomic Status, Parenting, and Child Development*, ed. M. H. Bornstein and R. H. Bradley, pp. 147–160. Mahwah, NJ: Lawrence Erlbaum Associates.

Hofstadter, D. R. 1979. *Gödel, Escher, Bach: An Eternal Golden Braid*. New York: Basic Books.

———. 2001. "Analogy as the Core of Cognition." In *The Analogical Mind: Perspectives from Cognitive Science*, ed. D. Gentner, K. J. Holyoak, and B. N. Kokinov, pp. 499–538. Cambridge, MA: MIT Press.

Hofstadter, D. R., and D. C. Dennett, eds. 1981. *The Mind's I*. New York: Basic Books.

Homer. 2004. *The Iliad*, trans. Robert Fitzgerald. New York: Farrar, Straus and Giroux.

———. 1998. *The Odyssey*, trans. Robert Fitzgerald. New York: Farrar, Straus, and Giroux.

Howson, C., and P. Urbach. 1991. "Bayesian Reasoning in Science." *Nature* 350: 371–374.

Hume, D. 1740. *A Treatise of Human Nature*. Available at: http://www.gutenberg.org/etext/4705.

Hunt, G. R., and R. D. Gray. 2003. "Diversification and Cumulative Evolution in New Caledonian Crow Tool Manufacture." *Proceedings of the Royal Society of London B* 270: 867–874.

Iglói, K., C. F. Doeller, A. Berthoz, L. Rondi-Reig, and N. Burgess. 2010. "Lateralized Human Hippocampal Activity Predicts Navigation Based on Sequence or Place Memory." *Proceedings of the National Academy of Sciences* 107: 14466–14471.

Irwin, T. H. 1991. "The Structure of Aristotelian Happiness." *Ethics* 101: 382–391. Review of *Aristotle on the Human Good* (1990) by Richard Kraut.

Isen, A. M. 2001. "An Influence of Positive Affect on Decision Making in Complex Situations: Theoretical Issues with Practical Implications." *Journal of Consumer Psychology* 11: 75–85.

———. 2008. "Some Ways in Which Positive Affect Influences Decision Making and Problem Solving." In *Handbook of Emotions*, 3rd ed., ed. M. Lewis, J. Haviland-Jones, and L. F. Barrett, pp. 548–573. New York: Guilford.

Jablonka, E., and M. Lamb. 2005. *Evolution in Four Dimensions*. Cambridge, MA: MIT Press.

———. 2007. "Evolution in Four Dimensions" (book précis). *Behavioral and Brain Sciences* 30: 353–392.

James, W. 1890. *The Principles of Psychology*. New York: Holt. Available at: http://psychclassics.yorku.ca/James/Principles/.

———. 1912. *Essays in Radical Empiricism*. New York: Longmans, Green and Co. Available at: http://www.brocku.ca/MeadProject/James/James_1912/James_1912_toc.html.

Johnson, S. 1765. *Preface to Shakespeare by Samuel Johnson*. Available online at Project Gutenberg, http://www.gutenberg.org/catalog/world/readfile?fk_files=1459392.

Jungman, N., A. Levi, A. Aperman, and S. Edelman. 1994. "Automatic Classification of Police Mug-Shot Album Using Principal Component Analysis." In *Proceedings of the SPIE-2243 Conference on Applications of Artificial Neural Networks* (Orlando, FL), ed. S. K. Rogers and D. W. Ruck, pp. 591–594.

Kahneman, D., and A. Deaton. 2010. "High Income Improves Evaluation of Life but Not Emotional Well-being." *Proceedings of the National Academy of Sciences* 107: 16489–16493.

Kalansuriya, A. D. P. 1993. "The Buddha and Wittgenstein: A Brief Philosophical Exegesis." *Asian Philosophy* 3: 103–112.

Kendrick, K. M., K. Atkins, M. R. Hinton, P. Heavens, and B. Keverne. 1996. "Are Faces Special for Sheep? Evidence from Facial and Object Discrimination Learning Tests Showing Effects of Inversion and Social Familiarity." *Behavioral Processes* 38: 19–35.

Kendrick, K. M., and B. A. Baldwin. 1987. "Cells in Temporal Cortex of Conscious Sheep Can Respond Preferentially to the Sight of Faces." *Science* 236: 448–450.

Killingsworth, M. A., and D. T. Gilbert. 2010. "A Wandering Mind Is an Unhappy Mind." *Science* 330: 932.

Kimchi, T., M. Reshef, and J. Terkel. 2005. "Evidence for the Use of Reflected Self-Generated Seismic Waves for Spatial Orientation in a Blind Subterranean Mammal." *Journal of Experimental Biology* 208: 647–659.

Knill, D., and W. Richards, eds. 1996. *Perception as Bayesian Inference*. Cambridge: Cambridge University Press.

Körding, K. P., and D. M. Wolpert. 2006. "Bayesian Decision Theory in Sensorimotor Control." *Trends in Cognitive Sciences* 10: 319–326.

Kringelbach, M. L., and K. C. Berridge. 2009. "Towards a Functional Neuroanatomy of Pleasure and Happiness." *Trends in Cognitive Sciences* 13: 479–487.

Kulish, N. 2009. "Spy Fired Shot That Changed West Germany." *New York Times*, May 27, p. A4.

Kushnir, T., and A. Gopnik. 2005. "Young Children Infer Causal Strength from Probabilities and Interventions." *Psychological Science* 16: 678–683.

Labroo, A. A., and S. Kim. 2009. "The 'Instrumentality' Heuristic: Why Metacognitive Difficulty Is Desirable During Goal Pursuit." *Psychological Science* 20: 127–134.

Lando, M. and S. Edelman. 1995. "Receptive Field Spaces and Class-Based Generalization from a Single View in Face Recognition." *Network* 6: 551–576.

Lao Tze. 1904. *The Book of the Simple Way of Laotze*, trans. W. G. Old. London: Philip Wellby.

Laozi. *Dao De Jing*, trans. James Legge. http://ctext.org/dao-de-jing.

Larkin, P., ed. 1973. *The Oxford Book of Twentieth-Century English Verse*. Oxford: Oxford University Press.

Larsen, J. T., and A. R. McKibban. 2008. "Is Happiness Having What You Want, Wanting What You Have, or Both?" *Psychological Science* 19: 371–377.

Lashley, K. S. 1951. "The Problem of Serial Order in Behavior." In *Cerebral Mechanisms in Behavior*, ed. L. A. Jeffress, pp. 112–146. New York: Wiley.

Lefebvre, L., S. M. Reader, and D. Sol. 2004. "Brain, Innovation and Evolution in Birds and Primates." *Brain, Behavior, and Evolution* 63: 233–246.

Lefebvre, L., and D. Sol. 2008. "Brains, Lifestyles, and Cognition: Are There General Trends?" *Brain, Behavior, and Evolution* 72: 135–144.

Le Guin, U. K. 1985. *Always Coming Home*. New York: Harper & Row.

———. 1968. *A Wizard of Earthsea*. Neptune, NJ: Parnassus Press.

Leising, K. J., J. Wong, M. R. Waldmann, and A. P. Blaisdell. 2008. "The Special Status of Actions in Causal Reasoning in Rats." *Journal of Experimental Psychology: General* 137: 514–527.

Lem, S. 1999. *A Perfect Vacuum*, trans. M. Kandel. Chicago: Northwestern University Press. Originally published in 1971.

Lindsay, P. H., and D. A. Norman. 1972. *Human Information Processing: An Introduction to Psychology*. New York: Academic Press.

Lyubomirsky, S., K. M. Sheldon, and D. Schkade. 2005. "Pursuing Happiness: The Architecture of Sustainable Change." *Review of General Psychology* 9: 111–131.

Mach, E. 1886. *Contributions to the Analysis of the Sensations*. New York: Open Court.

Maguire, E. A., D. G. Gadian, I. S. Johnsrude, C. D. Good, J. Ashburner, R. S. J. Frackowiak, and C. D. Frith. 2000. "Navigation-Related Structural Change in the Hippocampi of Taxi Drivers." *Proceedings of the National Academy of Sciences* 97: 4398–4403.

Mameli, M., and P. Bateson. 2006. "Innateness and the Sciences." *Biology and Philosophy* 21: 155–188.

Manuel, F. E., and F. P. Manuel. 1972. "Sketch for a Natural History of Paradise." *Daedalus* 101, no. 1 ("Myth, Symbol, and Culture"): 83–128.

Margrett, J., P. Martin, J. L. Woodard, L. S. Miller, M. MacDonald, J. Baenziger, I. C. Siegler, A. Davey, and L. Poon. 2010. "Depression Among Centenarians and the Oldest Old: Contributions of Cognition and Personality." *Gerontology* 56: 93–99.

Markman, E. 1989. *Categorization and Naming in Children*. Cambridge, MA: MIT Press.

Maron, J. L. 1982. "Shell-Dropping Behavior of Western Gulls (*Larus occidentalis*)." *The Auk* 99: 565–569.

Marr, D. 1970. "A Theory for Cerebral Neocortex." *Proceedings of the Royal Society of London B* 176: 161–234.

Marr, D., and T. Poggio. 1977. "From Understanding Computation to Understanding Neural Circuitry." *Neurosciences Research Program Bulletin* 15: 470–488.

Mason, M. F., M. I. Norton, J. D. van Horn, D. M. Wegner, S. T. Grafton, and C. N. Macrae. 2007. "Wandering Minds: The Default Network and Stimulus-Independent Thought." *Science* 315: 393–395.

Mayer, U., S. Watanabe, and H.-J. Bischof. 2010. "Hippocampal Activation of Immediate Early Genes Zenk and c-Fos in Zebra Finches (*Taeniopygia guttata*) During Learning and Recall of a Spatial Memory Task." *Neurobiology of Learning and Memory* 93: 322–329.

McCulloch, W. S. 1965. *Embodiments of Mind*. Cambridge, MA: MIT Press.

McDermott, D. V. 2001. *Mind and Mechanism*. Cambridge, MA: MIT Press.

Meilinger, T., M. Knauff, and H. H. Bülthoff. 2008. "Working Memory in Way-finding: A Dual Task Experiment in a Virtual City." *Cognitive Science* 32: 755–770.

Merker, B. 1993. "Vehicles of Hope: Hidden Structures in Man's Great Religions and Ideologies." Unpublished manuscript. Available at: http://www.pathsplitter.net/toc.php?visa=1.3.2.

———. 2004. "Cortex, Countercurrent Context, and Dimensional Integration of Lifetime Memory." *Cortex* 40: 559–576.

———. 2007. "Consciousness Without a Cerebral Cortex: A Challenge for Neuroscience and Medicine." *Behavioral and Brain Sciences* 30: 63–81.

Metzinger, T. 2003. *Being No One: The Self-Model Theory of Subjectivity*. Cambridge, MA: MIT Press.

Minsky, M. 1985. *The Society of Mind*. New York: Simon & Schuster.

———. 2006. *The Emotion Machine: Commonsense Thinking, Artificial Intelligence, and the Future of the Human Mind*. New York: Simon & Schuster.

Moore, J. W., D. Lagnado, D. C. Deal, and P. Haggard. 2009. "Feelings of Control: Contingency Determines Experience of Action." *Cognition* 110: 279–283.

Morris, W. 1971. *The Water of the Wondrous Isles*. New York: Ballantine Books.

Moses, Y., S. Ullman, and S. Edelman. 1996. "Generalization to Novel Images in Upright and Inverted Faces." *Perception* 25: 443–462.

Nagel, T. 1972. "Aristotle on Eudaimonia." *Phronesis* 17: 252–259.

———. 1974. "What Is It Like to Be a Bat?" *Philosophical Review* 83: 435–450. Available at: http://organizations.utep.edu/Portals/1475/nagel_bat.pdf.

Nicolao, L., J. R. Irwin, and J. K. Goodman. 2009. "Happiness for Sale: Do Experiential Purchases Make Consumers Happier Than Material Purchases?" *Journal of Consumer Research* 36: 188–198.

Niedenthal, P. M., L. W. Barsalou, P. Winkielman, S. Krauth-Gruber, and F. Ric. 2005. "Embodiment in Attitudes, Social Perception, and Emotion." *Personality and Social Psychology Review* 9: 184–211.

Noë, A. 2004. *Action in Perception*. Cambridge, MA: MIT Press.

O'Keefe, J., and L. Nadel. 1978. *The Hippocampus as a Cognitive Map*. Oxford: Clarendon Press.

Onnis, L., H. R. Waterfall, and S. Edelman. 2008. "Learn Locally, Act Globally: Learning Language from Variation Set Cues." *Cognition* 109: 423–430.

O'Regan, J. K., and A. Noë. 2001. "A Sensorimotor Account of Vision and Visual Consciousness." *Behavioral and Brain Sciences* 24: 883–917.

O'Toole, A. J., and S. Edelman. 1996. "Face Distinctiveness in Recognition Across Viewpoint: An Analysis of the Statistical Structure of Face Spaces." In *Proceedings of the Second International Conference on Automatic Face and Gesture Recognition*, ed. I. Essa, pp. 10–15. Washington, DC: IEEE Computer Society.

O'Toole, A. J., S. Edelman, and H. H. Bülthoff. 1998. "Stimulus-Specific Effects in Face Recognition over Changes in Viewpoint." *Vision Research* 38: 2351–2363.

Oxley, D. R., K. B. Smith, J. R. Alford, M. V. Hibbing, J. L. Miller, M. Scalora, P. K. Hatemi, and J. R. Hibbing. 2008. "Political Attitudes Vary with Physiological Traits." *Science* 321: 1667–1670.

Paulson, J. F., H. A. Keefe, and J. A. Leiferman. 2009. "Early Parental Depression and Child Language Development." *Journal of Child Psychology and Psychiatry* 50: 254–262.

Paz, R., H. Gelbard-Sagiv, R. Mukamel, M. Harel, R. Malacha, and I. Fried. 2010. "A Neural Substrate in the Human Hippocampus for Linking Successive Events." *Proceedings of the National Academy of Sciences* 107: 6046–6051.

Peirce, C. S. 1868. "Some Consequences of Four Incapacities." *Journal of Speculative Philosophy* 2: 140–157.

Philipona, D., J. K. O'Regan, J.-P. Nadal, and O. J.-M. Coenen. 2004. "Perception of the Structure of the Physical World Using Unknown Sensors and Effectors." *Advances in Neural Information Processing Systems* 15: 945–952.

Pinker, S. 2010. "The Cognitive Niche: Coevolution of Intelligence, Sociality, and Language." *Proceedings of the National Academy of Sciences* 107, suppl. 2: 8993–8999.

Proffitt, D. R. 2006. "Embodied Perception and the Economy of Action." *Perspectives on Psychological Science* 1: 110–122.

Quine, W. V. O. 1960. *Word and Object*. Cambridge, MA: MIT Press.

Raby, C. R., D. M. Alexis, A. Dickinson, and N. S. Clayton. 2007. "Planning for the Future by Western Scrub-Jays." *Nature* 445: 919–921.

Ramachandran, V. S., and W. Hirstein. 1997. "Three Laws of Qualia." *Journal of Consciousness Studies* 4: 429–458.

Reichard, W. 2007. "An Open Door." In *This Brightness*. Minneapolis: Mid-List Press.

Roberts, W. A., and M. C. Feeney. 2009. "The Comparative Study of Mental Time Travel." *Trends in Cognitive Sciences* 13: 271–277.

Rolls, E. T., G. C. Baylis, M. E. Hasselmo, and V. Nalwa. 1989. "The Effect of Learning on the Face Selective Responses of Neurons in the Cortex in the Superior Temporal Sulcus of the Monkey." *Experimental Brain Research* 76: 153–164.

Rumi, J., and S. Ahmed. 1881. *The Mesnavi and the Acts of the Adepts*, trans. J. W. Redhouse. London: Luzac & Co.

Russell, J., D. Alexis, and N. Clayton. 2010. "Episodic Future Thinking in Three- to Five-Year-Old Children: The Ability to Think of What Will Be Needed from a Different Point of View." *Cognition* 114: 56–71.

Saffran, J. R., R. N. Aslin, and E. L. Newport. 1996. "Statistical Learning by Eight-Month-Old Infants." *Science* 274: 1926–1928.

Saffran, J., and D. Wilson. 2003. "From Syllables to Syntax: Multilevel Statistical Learning by Twelve-Month-Old Infants." *Infancy* 4: 273–284.

Saunders, G. 2008. "Antiheroes," *The New Yorker*, June 23.

Scharfstein, B. 1998. *A Comparative History of World Philosophy: From the Upanishads to Kant*. Albany: State University of New York Press.

———. 2009. *Art Without Borders: A Philosophical Exploration of Art and Humanity*. Chicago: University of Chicago Press.

Schmidhuber, J. 2008. "Driven by Compression Progress: A Simple Principle Explains Essential Aspects of Subjective Beauty, Novelty, Surprise, Interestingness, Attention, Curiosity, Creativity, Art, Science, Music, Jokes." arXiv:0812.4360v1 [cs.AI], December 23.

Schoon, I., S. Parsons, R. Rush, and J. Law. 2010. "Children's Language Ability and Psychosocial Development: A Twenty-Nine-Year Follow-up Study." *Pediatrics* 126: e73–e80.

Scott-Phillips, T. C., and S. Kirby. 2010. "Language Evolution in the Laboratory." *Trends in Cognitive Sciences* 14: 411–417.

Sebeok, T., and J. U. Sebeok. 1981. "'You Know My Method': A Juxtaposition of Charles S. Peirce and Sherlock Holmes." In *The Play of Musement*, ed. T. Sebeok, pp. 17–52. Bloomington: Indiana University Press.

Senghas, A., S. Kita, and A. Özyürek. 2004. "Children Creating Core Properties of Language: Evidence from an Emerging Sign Language in Nicaragua." *Science* 305: 1779–1782.

Shams, L., and U. R. Beierholm. 2010. "Causal Inference in Perception." *Trends in Cognitive Sciences* 14: 425–432.

Siderits, M. 2007. *Buddhism as Philosophy*. Indianapolis, IN: Hackett.

Simon, H. A. 1973. "The Organization of Complex Systems." In *Hierarchy Theory: The Challenge of Complex Systems*, ed. H. H. Pattee, pp. 1–28. New York: George Braziller.

Sloman, A. 1989. "On Designing a Visual System (Towards a Gibsonian Computational Model of Vision)." *Journal of Experimental and Theoretical Artificial Intelligence* 1: 289–337.

Sloman, A., R. Chrisley, and M. Scheutz. 2005. "The Architectural Basis of Affective States and Processes." In *Who Needs Emotions? The Brain Meets the Robot*, ed. J. Fellous and M. A. Arbib, pp. 203–244. New York: Oxford University Press.

Smulders, T. V., A. D. Sasson, and T. J. DeVoogd. 1995. "Seasonal Variation in Hippocampal Volume in a Food-Storing Bird, the Black-Capped Chickadee." *Journal of Neurobiology* 27: 15–25.

Song, C., Z. Qu, N. Blumm, and A. Barabási. 2010. "Limits of Predictability in Human Mobility." *Science* 327: 1018–1021.

Speer, N. K., J. R. Reynolds, K. M. Swallow, and J. M. Zacks. 2009. "Reading Stories Activates Neural Representations of Visual and Motor Experiences." *Psychological Science* 20: 989–999.

Spivey, M. J. 2006. *The Continuity of Mind*. New York: Oxford University Press.

Stein, A., L. Malmberg, K. Sylva, J. Barnes, P. Leach, and the FCCC Team. 2008. "The Influence of Maternal Depression, Caregiving, and Socioeconomic Status in the Post-Natal Year on Children's Language Development." *Child: Care, Health and Development* 34: 603–612.

Stenning, K., and M. van Lambalgen. 2008. *Human Reasoning and Cognitive Science*. Cambridge, MA: MIT Press.

Stephens, G. J., L. J. Silbert, and U. Hasson. 2010. "Speaker-Listener Neural Coupling Underlies Successful Communication." *Proceedings of the National Academy of Sciences* 107: 14425–14430.

Steyvers, M., J. B. Tenenbaum, E. J. Wagenmakers, and B. Blum. 2003. "Inferring Causal Networks from Observations and Interventions." *Cognitive Science* 27: 453–489.

Stone, A. A., J. E. Schwartz, J. E. Broderick, and A. Deaton. 2010. "A Snapshot of the Age Distribution of Psychological Well-being in the United States." *Proceedings of the National Academy of Sciences* 107 (online May 17).

Strachan, J., ed. 2003. *A Routledge Literary Sourcebook on the Poems of John Keats*. London: Routledge.

Strugatsky, A. and B. Strugatsky. 1964. *Noon: XXII Century*. Moscow: Detgiz.

———. 1977. *Monday Starts on Saturday: A Fairy Tale for Younger Research Scientists*, trans. Leonid Renen. New York: DAW Books.

Swanwick, M. 1993. *The Iron Dragon's Daughter*. London: Millennium.

Taylor, A. H., G. R. Hunt, F. S. Medina, and R. D. Gray. 2009. "Do New Caledonian Crows Solve Physical Problems Through Causal Reasoning?" *Proceedings of the Royal Society of London B* 276: 247–254.

Tenenbaum, J. B., and T. L. Griffiths. 2001. "Generalization, Similarity, and Bayesian Inference." *Behavioral and Brain Sciences* 24: 629–641.

Tennyson, A. 1842. *Ulysses*. Available at: http://www.bartleby.com /246/375.html.

Tolkien, J. R. R. 1937. *The Hobbit*. London: George Allen & Unwin.

———. 1954. *The Lord of the Rings*, vol. 1, *The Fellowship of the Ring*. London: George Allen & Unwin.

Tolman, E. C. 1948. "Cognitive Maps in Rats and Men." *Psychological Review* 55: 189–208.

Turk-Browne, N. B., B. J. Scholl, M. M. Chun, and M. K. Johnson. 2009. "Neural Evidence of Statistical Learning: Efficient Detection of Visual Regularities Without Awareness." *Journal of Cognitive Neuroscience* 21: 1934–1945.

Van Boven, L., and T. Gilovich. 2003. "To Do or to Have: That Is the Question." *Journal of Personality and Social Psychology* 85: 1193–1202.

Veenhoven, R. 1994. "Is Happiness a Trait? Tests of the Theory That a Better Society Does Not Make People Any Happier." *Social Indicators Research* 32: 101–160.

Virgil. 29 B.C.E. *Georgics*. Available at: http://classics.mit.edu/Virgil /georgics.html.

Voltaire. 1924. *Voltaire's Philosophical Dictionary* (1752), trans. H. I. Woolf. New York: Knopf. Available at: http://history.hanover.edu /texts/voltaire/volindex.html.

Wardle, D., trans. 2006. *Cicero: On Divination*. New York: Oxford University Press.

Waterfall, H. R., and S. Edelman. 2009. "The Neglected Universals: Learnability Constraints and Discourse Cues" (a commentary on *Universals and Cultural Variation in Turn-Taking in Conversation* by Evans and Levinson). *Behavioral and Brain Sciences* 32: 471–472.

Waterfall, H. R., B. Sandbank, L. Onnis, and S. Edelman. 2010. "An Empirical Generative Framework for Computational Modeling of Language Acquisition." *Journal of Child Language* 37 (special issue 3): 671–703.

Watson, G. 2001. "Buddhism Meets Western Science." *Religion and the Brain* 19 (January).

Waytz, A., K. Gray, N. Epley, and D. M. Wegner. 2010. "Causes and Consequences of Mind Perception." *Trends in Cognitive Sciences* 14: 383–388.

Wegner, D. M. 2004. "The Illusion of Conscious Will" (précis). *Behavioral and Brain Sciences* 27: 649–692.

Wertsch, J. V. 1998. *Mind as Action*. Oxford: Oxford University Press.

Whitman, W. 1990. *Leaves of Grass: Song of the Open Road*, Oxford Classics edition. New York: Oxford University Press. Originally published in 1892.

Wilson, R. A. 1981. *Masks of the Illuminati*. New York: Dell.

Wittgenstein, L. 1958. *Philosophical Investigations*, trans. G. E. M. Anscombe, 3rd ed. Englewood Cliffs, NJ: Prentice-Hall.

Wood, E., P. A. Dudchenko, R. J. Robitsek, and H. Eichenbaum. 2000. "Hippocampal Neurons Encode Information About Different Types of Memory Episodes Occurring in the Same Location." *Neuron* 27: 623–633.

Young, L. J., and Z. Wang. 2004. "The Neurobiology of Pair Bonding." *Nature Neuroscience* 7: 1048–1054.

Young, M. P., and S. Yamane. 1992. "Sparse Population Coding of Faces in the Inferotemporal Cortex." *Science* 256: 1327–1331.

INDEX